Artificial Neural Networks for Renewable Energy Systems and Real-world Applications

Artificial Neural Networks for Renewable Energy Systems and Real-world Applications

Edited by

AMMAR H. ELSHEIKH

Production Engineering and Mechanical Design Department, Tanta University, Tanta, Egypt

MOHAMED ELASYED ABD ELAZIZ

Department of Mathematics, Faculty of Science, Zagazig University, Zagazig, Egypt

ELSEVIER

ACADEMIC PRESS

An imprint of Elsevier

Academic Press is an imprint of Elsevier
125 London Wall, London EC2Y 5AS, United Kingdom
525 B Street, Suite 1650, San Diego, CA 92101, United States
50 Hampshire Street, 5th Floor, Cambridge, MA 02139, United States
The Boulevard, Langford Lane, Kidlington, Oxford OX5 1GB, United Kingdom

ISBN: 978-0-12-820793-2

For Information on all Academic Press publications
visit our website at https://www.elsevier.com/books-and-journals

Publisher: Charlotte Cockle
Acquisitions Editor: Edward Payne
Editorial Project Manager: Sara Valentino
Production Project Manager: Anitha Sivaraj
Cover Designer: Mark Rogers

Typeset by MPS Limited, Chennai, India

Contents

List of contributors

Hayfa Y. Abuaddous
Department of Scientific Information and Services, Umm Al-Qura University, Mecca, Saudi Arabia

Laith Abualigah
Faculty of Computer Sciences and Informatics, Amman Arab University, Amman, Jordan; Hourani Center for Applied Scientific Research, Al-Ahliyya Amman University, Amman, Jordan; Faculty of Information Technology, Middle East University, Amman, Jordan; School of Computer Sciences, Universiti Sains Malaysia, Pulau Pinang, Malaysia; Department of Scientific Information and Services, Umm Al-Qura University, Mecca, Saudi Arabia

S. Agostinelli
Department of Astronautics, Electrical and Energy Engineering, Sapienza University of Rome, Rome, Italy

W. Ajbar
Centro de Investigación en Ingeniería y Ciencias Aplicadas (CIICAp), Universidad Autónoma del Estado de Morelos (UAEM), Cuernavaca, Morelos, México

Mohammed A.A. Al-qaness
State Key Laboratory for Information Engineering in Surveying, Mapping and Remote Sensing, Wuhan University, Wuhan, P.R. China

Mohammad Alshinwan
Department of Scientific Information and Services, Umm Al-Qura University, Mecca, Saudi Arabia

Leticia Chico-Santamarta
Harper Adams University, Newport, Shropshire, United Kingdom

Adriana Correa-Guimaraes
University of Valladolid, Department of Agricultural and Forestry Engineering, University of Valladolid, Palencia, Castile and León, Spain

Francisco J. Diez
University of Valladolid, Department of Agricultural and Forestry Engineering, University of Valladolid, Palencia, Castile and León, Spain

Youness El Mghouchi
Moulay Ismail University, National School of Arts and Crafts (ENSAM), Meknes, Morocco

Mohamed Elasyed Abd Elaziz
Department of Mathematics, Faculty of Science, Zagazig University, Zagazig, Egypt

Ammar H. Elsheikh
Production Engineering and Mechanical Design Department, Tanta University, Tanta, Egypt

Fadl A. Essa
Mechanical Engineering Department, Faculty of Engineering, Kafrelsheikh University, Kafrelsheikh, Egypt

Ismail Ghoul
Department of Computer Sciences and Applications, New Information and Communication Technologies Faculty, Laboratory of Complex Systems' Modeling and Implementation, Abdelhamid Mehri Constantine 2 University, Constantine, Algeria

J.A. Hernández
Centro de Investigación en Ingeniería y Ciencias Aplicadas (CIICAp), Universidad Autónoma del Estado de Morelos (UAEM), Cuernavaca, Morelos, México

A. Heydari
CITERA Interdepartmental Centre, Sapienza University of Rome, Rome, Italy

Rehab Ali Ibrahim
Department of Mathematics, Faculty of Science, Zagazig University, Zagazig, Egypt

Ahmad M. Khasawneh
Department of Scientific Information and Services, Umm Al-Qura University, Mecca, Saudi Arabia

Wenjia Li
Key Laboratory of Efficient Utilization of Low and Medium Grade Energy, MOE, Tianjin University, Tianjin, P.R. China

Andrés Martínez-Rodríguez
University of Valladolid, Department of Agricultural and Forestry Engineering, University of Valladolid, Palencia, Castile and León, Spain

Wafa Mousser
Department of Computer Sciences and Applications, New Information and Communication Technologies Faculty, Laboratory of Complex Systems' Modeling and Implementation, Abdelhamid Mehri Constantine 2 University, National Biotechnology Research Center, Constantine, Algeria

Luis M. Navas-Gracia
University of Valladolid, Department of Agricultural and Forestry Engineering, University of Valladolid, Palencia, Castile and León, Spain

Mahmoud Omari
Faculty of Computer Sciences and Informatics, Amman Arab University, Amman, Jordan; Department of Scientific Information and Services, Umm Al-Qura University, Mecca, Saudi Arabia

Mohammed Otair
Faculty of Computer Sciences and Informatics, Amman Arab University, Amman, Jordan

Salima Ouadfel
Department of Computer Sciences and Applications, New Information and Communication Technologies Faculty, Abdelhamid Mehri Constantine 2 University, Constantine, Algeria

A. Parrales
Centro de Investigación en Ingeniería y Ciencias Aplicadas (CIICAp), Universidad Autónoma del Estado de Morelos (UAEM), Cuernavaca, Morelos, México

E.D. Reyes-Téllez
Centro de Investigación en Ingeniería y Ciencias Aplicadas (CIICAp), Universidad Autónoma del Estado de Morelos (UAEM), Cuernavaca, Morelos, México

S. Sadek
Key Laboratory of Efficient Utilization of Low and Medium Grade Energy, MOE, Tianjin University, Tianjin, P.R. China

Mohd Khaled Yousef Shambour
Faculty of Information Technology, Middle East University, Amman, Jordan

S. Shanmugan
Research Centre for Solar Energy, Department of Engineering Physics, College of Engineering, Koneru Lakshmaiah Education Foundation, Guntur, Andra Pradesh, India

Mohammad Shehab
Information Technology, The World Islamic Sciences and Education University, Amman, Jordan; Faculty of Computer Sciences and Informatics, Amman Arab University, Amman, Jordan

Abdelmalik Taleb-Ahmed
Institut d'Electronique de Microélectronique et de Nanotechnologie (IEMN), UMR 8520, Université Polytechnique Hauts de France, Université de Lille, CNRS, Valenciennes, France

Somaya Zacout
Faculty of Computer Sciences and Informatics, Amman Arab University, Amman, Jordan

Mohamed E. Zayed
Key Laboratory of Efficient Utilization of Low and Medium Grade Energy, MOE, Tianjin University, Tianjin, P.R. China; Mechanical Power Engineering Department, Faculty of Engineering, Tanta University, Tanta, Egypt

Jun Zhao
Key Laboratory of Efficient Utilization of Low and Medium Grade Energy, MOE, Tianjin University, Tianjin, P.R. China

About the editors

Dr. Ammar H. Elsheikh received BS and MS degrees in mechanical engineering from Tanta University, Tanta, Egypt, and a PhD degree from Huazhong University of Science and Technology, Wuhan, China. He is currently working as an associative professor in Tanta University and Tokyo Institute of Technology. He is one of the top 2% of influential scholars, which includes the 100,000 top scientists in the world. His research interests include renewable energy, manufacturing processes, and the application of artificial intelligence techniques in engineering problems.

Dr. Mohamed Elasyed Abd Elaziz received BS and MS degrees in computer science from Zagazig University in 2008 and 2011, respectively. He received a PhD degree in mathematics and computer science from Zagazig University, Egypt, in 2014. From 2008 to 2011, he was working as an assistant lecturer in the Department of Computer Science. He is a program director of artificial intelligence science at Galala University, Egypt. He is the author of more than 200 articles. Abd Elaziz is one of the top 2% of influential scholars, which includes the 100,000 top scientists in the world. His research interests include metaheuristic technique, security IoT, cloud computing machine learning, signal processing, image processing, and evolutionary algorithms.

CHAPTER ONE

Basics of artificial neural networks

Rehab Ali Ibrahim[1], Ammar H. Elsheikh[2], Mohamed Elasyed Abd Elaziz[1] and Mohammed A.A. Al-qaness[3]
[1]Department of Mathematics, Faculty of Science, Zagazig University, Zagazig, Egypt
[2]Production Engineering and Mechanical Design Department, Tanta University, Tanta, Egypt
[3]State Key Laboratory for Information Engineering in Surveying, Mapping and Remote Sensing, Wuhan University, Wuhan, P.R. China

Contents

1.1 Artificial neural networks

Artificial neural networks (ANNs) are widely distributed processors made up of basic processing units called neurons [1]. They have a built-in capability for storing experimental knowledge that is suitable for use. High-speed information processing, routing capabilities, fault tolerance, adaptiveness, generalization, and robustness are all excellent characteristics of ANNs. These features make ANNs useful tools for modeling, optimizing, and predicting the performance of various engineering systems. As a result, they have being used to solve complex nonlinear engineering problems in a number of real-world applications with acceptable cost and efficient computing time [2–9].

Artificial Neural Networks for Renewable Energy Systems and Real-World Applications.
DOI: https://doi.org/10.1016/B978-0-12-820793-2.00002-1

The neuron model used in many ANN models is made up of a series of links called synapses, each with its own weight, as shown Fig. 1.1. Each weight is multiplied by the x_j input. Thereafter, all weighted inputs are summed and an externally applied bias b_k is applied to lower or increase the summation's output v_k.

Also, the activation function $\varphi(\bullet)$ is employed to the output for decreasing the amplitude range for the output signal y_k into a finite value. These sequences are formulated as follows:

$$y_k = \varphi\left(\sum_{j=1}^{m}\left(w_{kj}\right) + b_k\right) \tag{1.1}$$

where k represents a neuron, and j represents a synapse number.

1.2 Types of neural networks

In this section, we describe four ANN models, including the multilayer perceptron (MLP), wavelet neural network (WNN), radial basis function (RBF), and Elman neural network (ENN).

1.2.1 Multilayer perceptron neural network

The following are the fundamental principles of the MLP neural network. In general, this form of ANN has one input layer, multiple hidden layers, and one output layer [10,11], as shown in Fig. 1.2A. The input neuron is

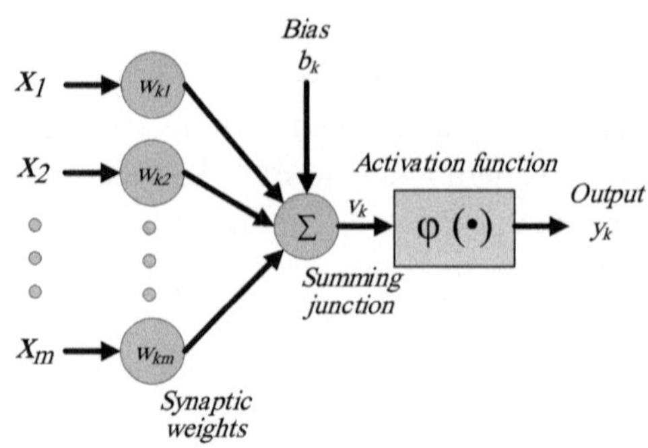

Figure 1.1 Nonlinear model of a neuron [1].

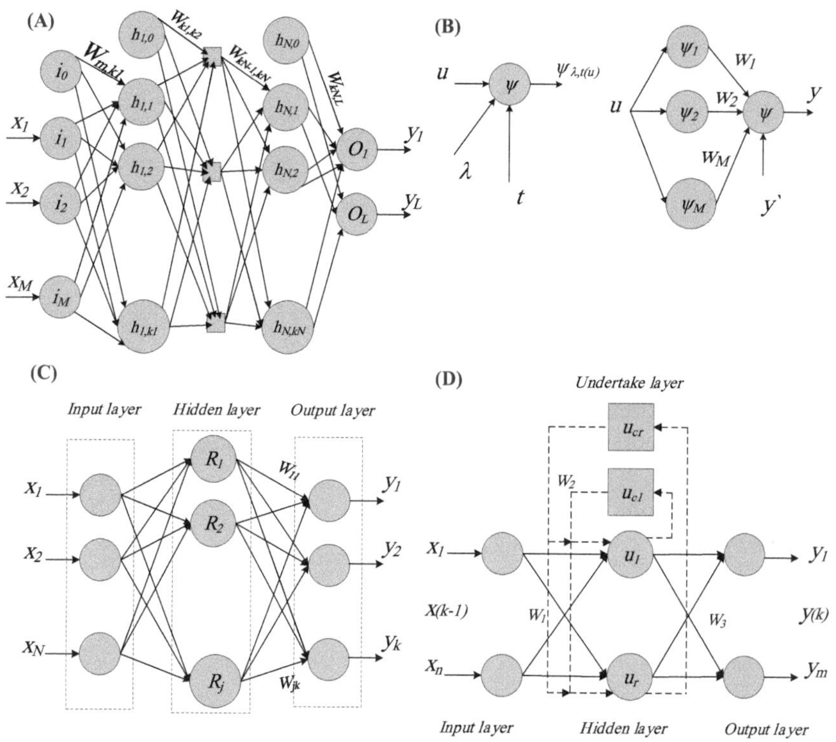

Figure 1.2 The structure of (A) MLP [12], (B) wavelet neural networks [13], (C) RBF [14], and (D) Elman networks [15].

connected to the hidden layer neuron, and the hidden layer neuron is connected to the output neuron, but the neurons of the same layer are not connected. The neurons in the input layer obtain the data and transfer them on to the next layers before reaching the output layer.

This problem can be defined as follows. There are M neurons in the input layer, and a k_1-th neuron receives the x_i input, where the output of the O_{1,k_1} neuron can be obtained by:

$$O_{1,k_1} = f\left(\sum_{m=0}^{M} w_{mk_1} x_m\right), k_1 = 1, \ldots, K_1 \tag{1.2}$$

Thus, the O_{1,k_1} output is utilized as an input to the next hidden layer. The output of the neuron in the hidden layers is calculated as:

$$O_{h,k_h} = f\left(\sum_{K_{h-1}=0}^{K_{h-1}} w_{k_{h-1},k_h} O_{h-1,k_{h-1}}\right), k_h = 2, \ldots, K_h, h = 1, \ldots, N_h \quad (1.3)$$

where $f(.)$ denotes the activation function, and $w_{0,k_h}, k_h = 1, \ldots, K_h$ are the biases of the input and hidden layers. Also, $w_{k_m,k_h}(w_{k_{h-1},k_h})$ denote the weights between the input and hidden layer neurons. M is the number of inputs, and K_h is the number of h hth hidden layer neurons, where N_h represents the number of hidden layers.

An output neuron can be considered as a weighted sum of all outputs of the neurons of the last hidden layer, as follows:

$$y_l = \sum_{k_N=0}^{K_N} w_{k_N,l} O_{N,k_N}, l = 1, \ldots, L \quad (1.4)$$

where L represents the number of neurons in the output layer, and $w_{kh,l} w_{0,k_h}, k_h = 1, \ldots, K_h$ represents the weights between neurons in hidden layers and the neurons in the output layer.

Since the weights' value has an effect on the final performance, back-propagation (BP) learning [16] is used to find these values. The BP method, on the other hand, takes a long time to complete because it requires an iterative training phase. Furthermore, the BP is most likely stuck in local minima.

1.2.2 Wavelet neural networks

WNNs, also known as wavelet networks (WNs), are a combination of wavelet theory and neural networks (NNs) [13]. The advantages of both the neural network and the wavelet transformation are inherited by WNNs. A WNN is made up of a feed-forward neural network and a hidden layer, the activation functions of which can be created using the orthonormal wavelet family [17]. The wavelet neurons are also called wavelons.

As shown in Fig. 1.2B, the WNN's simplest structure consists of an input layer and an output layer. The concealed layer of the WNN is made up of wavelons, with the wavelet dilation and translation parameters as input coefficients. Since the input is in a small area of the input space, the hidden layer's wavelons result in an output that is not zero. A WNN's output can be interpreted as a linear combination of the wavelet's activation functions, and this combination is also weighted.

Thus, the output can be formulated as the following equation:

$$\psi_{\lambda,t}(u) = \psi\left(\frac{u-t}{\lambda}\right) \tag{1.5}$$

where t denotes the translation, and λ denotes the dilation coefficients.

Fig. 1.2B portrays the general structure of a WNN with one input and one output. The WNN's secret layer is made up of wavelons with the number M. The network's output is a weighted sum of the wavelon's outputs [18].

$$y(u) = \sum_{i=1}^{M} w_i \psi_{\lambda_i,t_i}(u) + \overline{y} \tag{1.6}$$

where the $\overline{y}$ value deals with the functions that contain nonzero mean, and the wavelet function $\psi(u)$ has a mean with zero value. Also, at the highest scale, the value of $\overline{y}$ is considered instead of the function $\phi(u)$. In the WNN, $\overline{y}$, w_i, t_i, λ_i coefficients are learned by an algorithm.

1.2.3 Radial basis function

The RBF neural network is one of the most efficient NNs, with three layers: input, hidden, and output. It is similar to MLP, as illustrated in Fig. 1.2C [14,19].

The RBF's input layer receives the input data and only passes it through the hidden layer. Following that, the hidden layer processes the obtained data and extracts relevant information before sending it to the output layer, which constructs the output data. However, the most important distinction between RBF and MLP is that tRBF is used by the hidden layer, which is defined by the Gaussian function, which is formulated as [14]:

$$a_j(x) = \exp\left(\frac{x_i - c_j}{2\sigma_j^2}\right) \tag{1.7}$$

where σ_j denotes the width of the neuron with a number j-th. x_i represents the RBF input, and c_j is the center of the RBF unit. Also, there are several functions that can be applied, such as:

$$a_j(x) = \exp\left(-\left(\frac{x}{\sigma}\right)^2\right) \tag{1.8}$$

$$a_j(x) = \exp\left(\frac{1}{(\sigma^2 + x^2)^\alpha}\right), \alpha > 0 \tag{1.9}$$

$$a_j(x) = \left(\alpha^2 + x^2\right)^\beta, \alpha < \beta < 1 \tag{1.10}$$

Furthermore, another difference is the output layer using a linear function which is represented as the weighted sum of RBFs. This can be represented as:

$$y_k = \sum_{j=1}^{J} w_{jk} R_j(x), k = 1, 2..., K \tag{1.11}$$

where K is the outputs' number, and w_{jk} denotes the weight that connects the j-th node of the hidden layer and the k-th node of the output layer. Finally, j represents the number of nodes in the hidden layer.

1.2.4 Elman neural network

Generally, the ENN is a recurrent network (RNN), and a type of ANN in which connections between neurons form a guided cycle, which leads to dynamic temporal activity. RNNs vary from feedforward neural networks in that they handle input sequences using internal memory rather than external memory. This enables them to perform tasks such as unsegmented handwriting recognition and speech recognition.

Elman [20] proposed the ENNs, where the output of hidden layers is fed back onto themselves using the recurrent layer, and the number of recurrent neurons is equal to the hidden neurons. Therefore, ENNs have high ability for learning and constructing temporal and spatial patterns. Also, they have dynamic function that can give a system better capability to deal with time-varying properties.

The node in the hidden layer is connected only with one recurrent node using a fixed weight value.

The structure of ENNs is shown in Fig. 1.2D. They have four layers (input, hidden, recurrent, and output). Let the input and the output layers consist of N and M, and the hidden neurons number be denoted by N_h.

Table 1.1 Various activation functions.

Type	Description	Formula	Graph
Linear	The simplest activation function in which the output is proportional to the input	$\varphi(v_k) = cv_k$ in which c is the slope of the output of the function	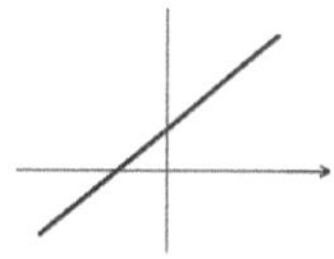
Sigmoid	A family of S-shaped functions	$\varphi(v_k) = \frac{1}{1 + e^{-v_k}}$	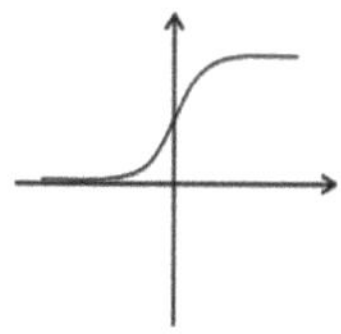
Gaussian	This is a continuous bell-shaped curve that has a peak at $v_k = 0$ and diminishes at the tails. The output node (high/low) is interpreted according to the membership of the class (1/0), based on how close the input of the net is to a value of the average which has been selected	$\varphi(v_k) = e^{-v_k^2}$	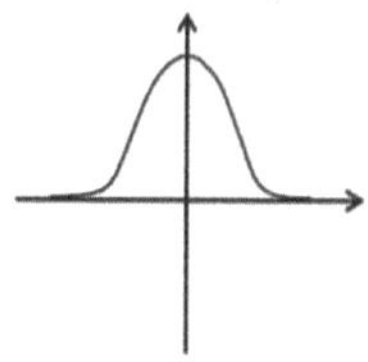
Gaussian complement	This is the inversion version of the Gaussian function that is continuous, diminishes at $v_k = 0$, and has peak at the tails	$\varphi(v_k) = 1 - e^{-v_k^2}$	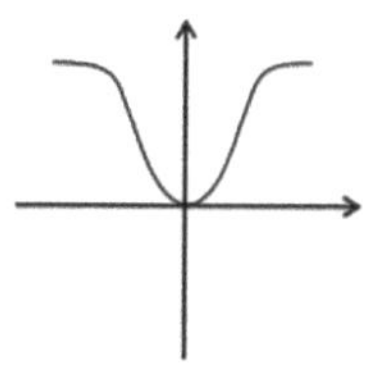

$u(k-1)$ is the input of the neural network, where $x(k)$ and $x_c(k)$ are the outputs of the hidden and recurrent layers, respectively, as defined in the following equation:

$$x(k) = f(w_2 x_c(k) + w_1 u(k-1)),\, x_c(k) = x(k-1) \tag{1.12}$$

where w_1 represents the weight of the connection between the nodes in the input and hidden layers. w_2 represents the weight of the connection between the hidden and recurrent layers, and f denotes the transfer function and is represented by the sigmoid function (as described in Table 1.1).

After that, the hidden layer output is calculated, and the output of the neural network $y(k)$ is calculated as:

$$y(k) = g(w_3 x(k)) \tag{1.13}$$

where w_3 denotes the weight between the hidden and output layers, and g represents the transfer function of the output layer.

The back-propagation is employed in ENNs for the weights updating process, furthermore, the error of the network is calculated as [15]:

$$E = \sum_{k=1}^{K} (d(k) - y(k))^2 \tag{1.14}$$

where $d(k)$ denotes the desired output of the $u(k)$ input.

1.2.5 Statistical performance evaluation criteria

To assess the quality of the prediction of NNs, a set of evaluation metrics are applied. These metrics are mean square error (*MSE*), mean absolute error (*MAE*), mean relative error (*MRE*), and root mean square error (RMSE). In addition there are the correlation coefficient (*Corr*), coefficient of variance (*Cov*), efficiency coefficient (*EC*), coefficient of determination (R^2), coefficient of residual mass (*CRM*), and through the index of the performance of the model (*OI*). These measures are defined in Table 1.2, in which n_s is the number of observations, d is the observed values, and y is the desired and prediction values. d_{max} and d_{min} are the maximum and the minimum observed values, respectively, and $\overline{y}$ and $\overline{d}$ represent the average of the observed and prediction values, respectively.

Table 1.2 Definition of performance measures.

Measure	Formula
Mean square error	$MSE = \frac{1}{n_s}\sum_{i=1}^{n_s}(d_i - y_i)^2$
Mean relative error	$MRE = \frac{1}{n_s}\sum_{i=1}^{n_s}\frac{d_i - y_i}{d_i}$
Mean absolute error	$MAE = \frac{1}{n_s}\sum_{i=1}^{n_s}\lvert d_i - y_i\rvert$
Root mean square error	$RMSE = \sqrt{\frac{1}{n_s}\sum_{i=1}^{n_s}(d_i - y_i)^2}$
Coefficient of variance	$Cov = \frac{RMSE}{\frac{1}{n_s}\sum_{i=1}^{n_s} y_i} \times 100$
Correlation coefficient	$Corr = \frac{n_s\sum_{i=1}^{n_s} d_i y_i - \left(\sum_{i=1}^{n_s} d_i\right)\left(\sum_{i=1}^{n_s} y_i\right)}{\left[n_s\sum_{i=1}^{n_s} d_i^2 - \left(\sum_{i=1}^{n_s} d_i\right)^2\right]\left[n_s\sum_{i=1}^{n_s} y_i^2 - \left(\sum_{i=1}^{n_s} y_i\right)^2\right]}$
Coefficient of determination	$R^2 = \frac{\left(\sum_{i=1}^{n_s}(d_i - \bar{d})(y_i - \bar{y})\right)^2}{\sum_{i=1}^{n_s}(d_i - \bar{d})^2 \times \sum_{i=1}^{n_s}(y_i - \bar{y})^2}$
Efficiency coefficient	$EC = 1 - \frac{\sum_{i=1}^{n_s}(d_i - y_i)^2}{\sum_{i=1}^{n_s}(d_i - \bar{d})^2}$
Overall index of model performance	$OI = \frac{1}{2}\left(1 - \left(\frac{RMSE}{d_{max} - d_{min}}\right) + EC\right)$
Coefficient of residual mass	$CRM = \frac{\sum_{i=1}^{n_s} y_i - \sum_{i=1}^{n_s} d_i}{\sum_{i=1}^{n_s} d_i}$

1.3 Conclusion

In this introductory chapter, different types of ANNs have been introduced. These ANN types include MLP, WNN, RBF, and ENN. The structure and mathematical notation of each ANN have been presented. It can be concluded that each of these ANNs has its own advantages that make it suitable for specific applications. In addition, most criteria that are used to assess the quality of the different ANN models have been presented.

References

[1] S.S. Haykin, et al., Neural Networks and Learning Machines, Vol. 3, Pearson, Upper Saddle River, NJ, 2009.
[2] M. Abd Elaziz, et al., Utilization of Random Vector Functional Link integrated with Marine Predators Algorithm for tensile behavior prediction of dissimilar friction stir welded aluminum alloy joints, Journal of Materials Research and Technology 9 (5) (2020) 11370–11381.
[3] H.A. Babikir, et al., Noise prediction of axial piston pump based on different valve materials using a modified artificial neural network model, Alexandria Engineering Journal (2019).
[4] M.A. Elaziz, A.H. Elsheikh, S.W. Sharshir, Improved prediction of oscillatory heat transfer coefficient for a thermoacoustic heat exchanger using modified adaptive neuro-fuzzy inference system, International Journal of Refrigeration (2019).
[5] E.M.S. El-Said, M. Abd Elaziz, A.H. Elsheikh, et al., Machine learning algorithms for improving the prediction of air injection effect on the thermohydraulic performance of shell and tube heat exchanger, Applied Thermal Engineering 185 (2021) 116471.
[6] Elsheikh, et al., Prediction of laser cutting parameters for polymethylmethacrylate sheets using random vector functional link network integrated with equilibrium optimizer, Journal of Intelligent Manufacturing (2020).
[7] A.H. Elsheikh, et al., A new artificial neural network model integrated with a cat swarm optimization algorithm for predicting the emitted noise during axial piston pump operation, IOP Conference Series: Materials Science and Engineering 973 (2020) 012035.
[8] A.H. Elsheikh, et al., Utilization of LSTM neural network for water production forecasting of a stepped solar still with a corrugated absorber plate, Process Safety and Environmental Protection 148 (2021) 273–282.
[9] A.H. Elsheikh, et al., Modeling of solar energy systems using artificial neural network: a comprehensive review, Solar Energy 180 (2019) 622–639.
[10] P.M. Atkinson, A. Tatnall, Introduction neural networks in remote sensing, International Journal of Remote Sensing 18 (4) (1997) 699–709.
[11] H. Yan, et al., A multilayer perceptron-based medical decision support system for heart disease diagnosis, Expert Systems with Applications 30 (2) (2006) 272–281.
[12] Y. Ren, et al., Random vector functional link network for short-term electricity load demand forecasting, Information Sciences 367 (2016) 1078–1093.
[13] Q. Zhang, A. Benveniste, Wavelet networks, IEEE Transactions on Neural Networks 3 (6) (1992) 889–898.
[14] D.S. Broomhead, D. Lowe, Radial Basis Functions, Multi-Variable Functional Interpolation and Adaptive Networks, Royal Signals and Radar Establishment Malvern, United Kingdom, 1988.
[15] W. Jia, et al., Study on optimized Elman neural network classification algorithm based on PLS and CA, Computational Intelligence and Neuroscience 2014 (2014) 12.
[16] C.-N. Ko, C.-M. Lee, Short-term load forecasting using SVR (support vector regression)-based radial basis function neural network with dual extended Kalman filter, Energy 49 (2013) 413–422.
[17] Z. Zhang, Iterative algorithm of wavelet network learning from nonuniform data, Neurocomputing 72 (13) (2009) 2979–2999.
[18] G. Wang, L. Guo, H. Duan, Wavelet neural network using multiple wavelet functions in target threat assessment, The Scientific World Journal 2013 (2013).
[19] D. Wu, et al., Prediction of Parkinson's disease tremor onset using radial basis function neural networks, Expert Systems with Applications 37 (4) (2010) 2923–2928.
[20] J.L. Elman, Finding structure in time, Cognitive Science 14 (2) (1990) 179–211.

CHAPTER TWO

Artificial neural network applied to the renewable energy system performance

A. Parrales, E.D. Reyes-Téllez, W. Ajbar and J.A. Hernández
Centro de Investigación en Ingeniería y Ciencias Aplicadas (CIICAp), Universidad Autónoma del Estado de Morelos (UAEM), Cuernavaca, Morelos, México

Contents

Nomenclature

ANN	artificial neural network
BR	Bayesian regularization training method
b_1	bias in the hidden layer
b_2	bias in the output layer
F	test F fisher
I_P	importance of the input variable
LM	Levenberg–Marquardt training method
MAPE	mean absolute percent error
RMSE	root mean square error
R^2	coefficient of determination
S	standard deviation combined
T	temperature (°C)
t	student's t-test
W_i	weights in the input layer
W_o	weights in the output layer
x	value

Artificial Neural Networks for Renewable Energy Systems and Real-World Applications.
DOI: https://doi.org/10.1016/B978-0-12-820793-2.00006-9

Greek letters

δ LOGSIG function
λ PURELIN function
τ TANSIG function

Subscripts

env ambient
exp experimental
glass glass surface
input fluid inlet
norm normalized
output fluid outlet
reflect reflective surface
sim simulated

2.1 Introduction

Since ancient times, energy has been the fundamental input for the development of human activities. With population growth, energy consumption has reached an excessive magnitude that needs to be satisfied. Therefore, currently, the technological advancement and economic growth of nations depend mainly on the adequate transformation and use of the energy sources provided by the environment: renewable and nonrenewable [1].

In recent years, most of the energy consumed in the world is nonrenewable; that is, once consumed, there is no way to recover it. Examples of sources of nonrenewables are nuclear energy and the burning of fossil fuel such as oil, coal, and natural gas, among others. With the increase in the use of fossil fuels, caused by the accelerated development of industry, negative consequences have been generated in the environment. As an example, the release of carbon dioxide into the atmosphere has induced high temperatures in some regions, water pollution, droughts, acid rains, and floods, among others [2].

According to data from the International Energy Agency, global carbon dioxide emissions grew 1.7% in 2018, reaching a record high of 33.1 Gt of CO_2; reported as the fastest growth since 2013 [3]. As a result, international cooperation has been sought to reduce climate change through 17 sustainable development goals set by the United Nations Organization (UN). Within these objectives, numbers 7 and 13 focused on increasing the use of renewable energies that contribute to reducing carbon dioxide emissions to maintain the global surface temperature increase at below 2°C [4]. Therefore, nations are developing technologies focused on obtaining renewable energy sources and replacing fossil fuels. With this, renewable energies are expected to represent two-thirds of the total global energy supply and thereby reduce carbon dioxide emissions into the atmosphere by 60% by 2050 [5].

Wahab et al. [6] concluded that the energy produced by renewable energy resources is the most sustainable, with a minimum emission of greenhouse gases. Renewable energies, also known as clean energies, are those that are provided by nature, not depleted by their use, and renewed continuously or over short periods, and include: hydraulic, solar, wind, tidal or marine (tide, waves, and ocean energy), and tree mass [7].

Currently, the share of renewable energies worldwide amounts to 2351 GW, equivalent to one-third of the total energy supply. As can be seen in Fig. 2.1, hydro represents the majority of this total with a 50% share (1172 GW) followed by wind power with 24% (564 GW). In third place is solar energy at 20% (486 GW) and finally the other energies together represent 6% of the total, including: bioenergy (115 GW), geothermal energy (13 GW), and marine energy (500 GW). According to data from the International Renewable Energy Agency, the expansion

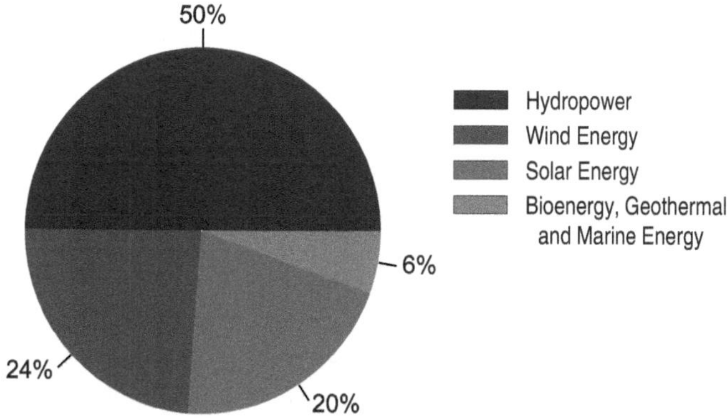

Figure 2.1 Percentages of renewable generation capacity by energy source in the year 2018.

of renewable capacity continues to be driven mainly by new solar and wind energy installations. In 2018, solar energy increased by 24% (171 GW), and wind energy increased by 10% (94 GW) compared to the previous year [8].

In particular, solar energy is the cleanest renewable energy source with great potential as a possible sustainable solution compared to any other energy source [9]. Furthermore, based on the estimates of NASA scientists, the Sun is less than half of its useful life, which means it will last 6500 million years until it is reduced to a white star dwarf [10]. Solar energy is obtained from the interactions in the Sun's core, where hydrogen is converted to helium and transmitted to Earth by electromagnetic waves or particles (photons) known as solar radiation [11].

The Sun produces an enormous amount of energy every day that reaches locations throughout the world, but solar radiation that crosses the atmosphere varies depending on factors such as geographical position, latitude, the season of the year, day, hour, and atmospheric conditions such as cloudiness [9]. Solar radiation provides two primary energies: light and heat that drive all environmental processes and that act directly or indirectly on the Earth's surface. Some examples are the photosynthesis of plants, the heated air masses that cause wind, the creation of fossil fuels over long periods, the evaporation of the oceans that causes rain and rivers providing hydroelectric energy, among other manifestations [11].

As expected, solar radiation has great potential for human consumption due to the annual flow of solar energy being almost 100 times greater than the annual global energy consumption. According to Rypkema [12], the collection of only 1% of the solar radiation incident could meet the energy needs of everyone on Earth. In so far as the methods of harnessing solar energy, there are two ways to collect it: either solar thermal energy or electrical energy, or their combination.

Electrical energy is obtained from solar photovoltaic (PV) cells, where the PV solar effect is achieved by direct conversion of sunlight into electricity without any heat engine to interfere [13]. PV systems are robust, require little maintenance, and are of simple design for applications such as power sources, water pumping, home lighting systems, desalination, and communications, among others [14]. However, the conversion efficiency of the electric energy produced by solar PV cells is very low, and it has been revealed that only 15%–20% of solar radiation is transformed into electricity [1].

On the other hand, solar thermal energy is captured by diverse types of solar thermal collectors (STCs). STCs are a particular type of heat exchanger used to absorb incoming solar energy, transform it into heat, and transfer it to a working fluid. The working fluid transports the collected solar energy either to be used or to a thermal energy storage tank [15]. At present, these systems can generate heat for domestic hot water purposes, zone heating, cooking (solar ovens), water treatments (desalination), and some industrial processes, among others [16]. As can be seen in Fig. 2.2, STCs are mainly classified into two types according to the concentration ratio: concentrated and nonconcentrated or flat technology. With the flat technology collectors, the maximum temperature levels obtained are generally up to 90°C–150°C, while the concentration technologies can reach temperatures of between 150°C and 500°C [17].

The simplest and most common nonconcentration solar collector design is the flat-plate collector. As the name implies, these collectors use a flat absorption surface to capture light using traditional insulating materials such as foam or fiberglass to retain heat. However, high thermal losses prevent these

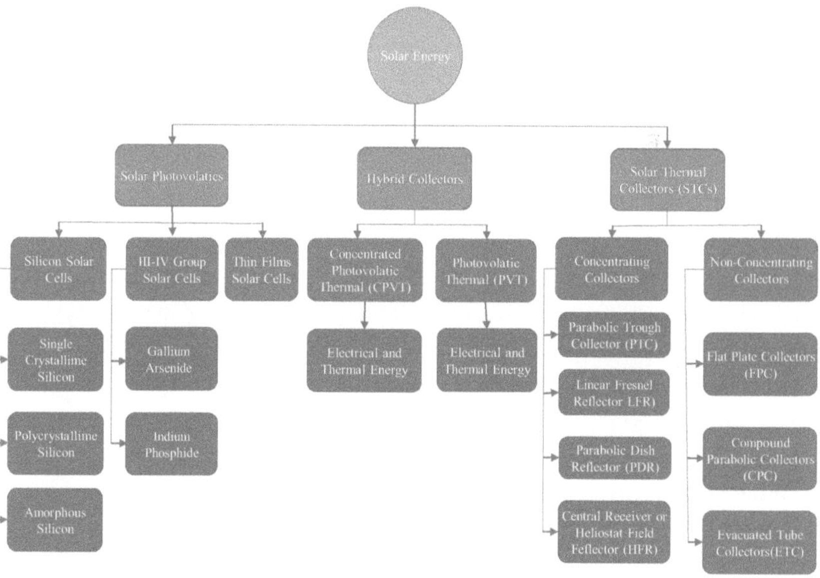

Figure 2.2 Classification of the diverse types of solar collectors.

systems from operating efficiently above 100°C [18]. Other designs that can be found are compound parabolic collectors and evacuated tube collectors.

With respect to concentrated solar collectors, they generally use mirrored surfaces to focus light from a large area onto a smaller area where the absorbent material is located. In comparison with nonconcentration systems, they directly absorb the light that reaches their surface [19]. The solar collectors classified as concentrates are: parabolic trough collectors, linear Fresnel reflectors, parabolic dish reflectors, and heliostat field reflectors.

In particular, the parabolic trough solar collector (PTC) stands out as a low-operating-cost technological system compared to other technologies used in solar concentration [20]. Also, PTCs have a wide range of thermal availability that allows meeting the heat energy requirements of some industrial processes or electric power generation. They have concentration ratios ranging from 15% to 45%, and can produce heat at temperatures of between 50°C and 400°C. Therefore they can be used in a great variety of applications such as cooling, industrial heating, and chemical processes, among others [21].

A PTC system is a technology composed by a parabolic trough reflecting layout used to concentrate the direct radiation in a focal line, where a receiver is placed. The design of the parabolic trough considers important parameters such as solar distribution flux, reflector material of the concentrator, diameter of the parabolic trough concentrator, size of the aperture area of the concentrator, focal length of the parabolic trough, the focal point diameter, size of the aperture area of the receiver, absorber material of the receiver, and geometric or area concentration ratio [22]. A scheme of a PTC is shown in Fig. 2.3.

The receiver is formed by an absorber tube and a glass pipe cover [23]. The absorber pipe is usually constructed from stainless steel coated by a selective surface for better absorption of solar radiation and to decrease emittance [24]. The annular space between the absorber and the glass envelope is in vacuum conditions to reduce thermal losses. Within the absorber tube, working fluid circulates to transport the thermal energy. There are seven main groups of working fluids: synthetic oils, molten salts, water/steam, air or pressurized gases, organics, liquid metals, and, recently, nanofluids [23].

The absorption of concentrated solar radiation within the receiver is the critical step that governs the overall thermal efficiency of the PTC system [25]. Numerous researches about solar collectors have been focused on how the different operating parameters affect thermal the performance, in order to improve it. However, experimental studies are constrained by

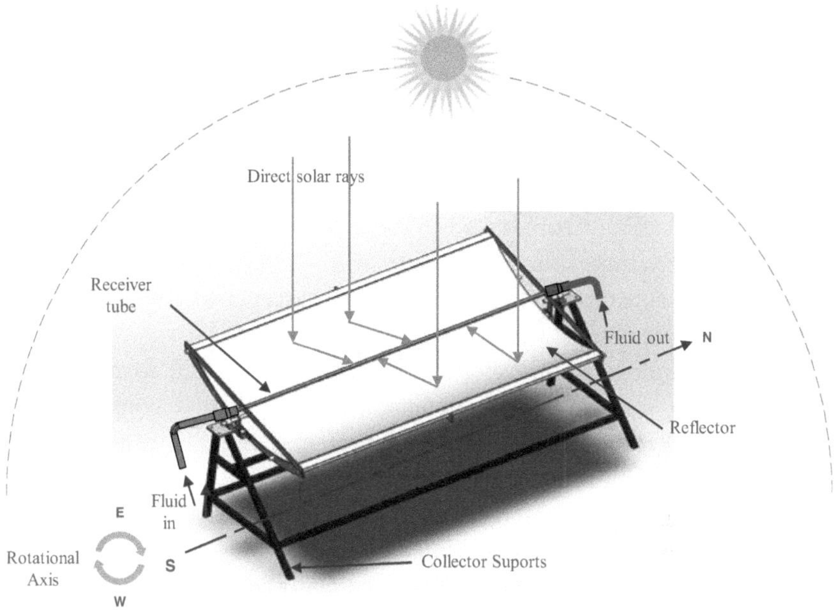

Figure 2.3 Schematic diagram of the parabolic trough solar collector system.

the number of conducted experiments which may require high operating cost and time [26]. Therefore, studies have been inclined toward the use of empirical models displaying the phenomenon accurately which and are able to replace the experiments.

One of the most used tools in empirical modeling of a nonlinear phenomenon has been artificial neural networks (ANNs) [27]. Specifically, the data obtained from renewable energy systems are good candidates to be managed with ANN, since it is difficult to identify a precise mathematical function when some of the parameters vary by time and environmental factors [28]. Furthermore, Kalogirou [29] suggested that models based on ANNs have demonstrated the effectiveness of their use in solar collector systems. ANNs have been used in areas of solar energy application such as forecasting of solar potential, solar desalination systems productivity, solar PV power output, among others. Also, the performance analysis of solar systems could be made simple and accurate with the use of ANNs to enhance the efficiency and productivity [30]. Some of these are mentioned below.

For example, in the study by Diez et al. [31] a model was developed based on ANNs to predict the outlet temperature of a flat-plate solar collector using measured data of solar irradiance, ambient temperature, inlet temperature, and working fluid flow. The results were acceptable compared to the experimental data obtained through statistical tests, which obtained coefficients of determination above 0.7443 for different working fluid flows.

Sözen et al. [32] developed a new formula based on ANNs to determine the efficiency of a flat-plate solar collector with experimental data that were measured from July to September, obtaining a coefficient of determination of 0.983. The variables used in the input layer of the network were: collector surface temperature, date, time, solar radiation, declination angle, azimuth angle, and tilt angle. The results were acceptable compared to the experimental data obtained through statistical tests.

An ANN model was developed by Cetiner et al. [33] to predict the efficiency of a solar hot water generator system, which consisted of a cylindrical concentrator, an absorber, a heat exchanger, a water store, a pump, and a control unit. The authors concluded that with the use of an ANN model useful energy, mass flow rate, and other parameters can be obtained without additional experiments.

With the use of ANNs, prediction of the performance parameters of flat-plate solar collectors has been realized by Kalogirou [29]. In this work, prediction of the performance coefficients, both in wind and no-wind conditions, the incidence angle, modifier coefficients at longitudinal and transverse directions, the collector time constant, the collector stagnation temperature, and the collector heat capacity were developed with six ANN models.

In another work, Heng et al. [34] used an ANN with the superposition principle for transient thermal prediction in a PTC. The ANN was used to predict the temperature increase at the PTC output for a single pulse. A backpropagation network was trained with a total of 29,884 samples, architecture of 3–68–68–1, and a gradient descent training algorithm obtaining an R^2 of 0.9999. The input variables were heated flow, fluid velocity, and time. The results were compared with the finite element method (FEM model), which had an $R^2 = 1$.

In the following section of this chapter, the development of an ANN neural model to predict the outlet temperature of a low-cost solar collector is described. This type of collector is specifically chosen as a possible option for use in residential areas since they can be manufactured with economical and feasible materials for installation in low-income communities. The materials used for its manufacture can be found in any region (copper tube and steel concentrator). In addition, it has a long service life due to its resistance, compared to other collectors using fragile materials. This ANN model considers essential parameters that influence the thermal behavior of the PTC.

The inlet temperature of the fluid, month, day, hour, ambient temperature, glass surface temperature, and reflective surface temperature were the input variables. Also, the calculated values were compared with experimental data to demonstrate the effectiveness of the capacity of the network.

2.2 Description of experimental equipment

The experimental study system was a static PTC which is mainly comprised of a reflective plate recessed in a parabolic concentrator, which has dimensions of 85 cm long by 57 cm in diameter. The reflective plate is made of 316L stainless steel 16-gauge type used to concentrate the Sun's rays along a focal line in which a receiving tube is placed.

The receiver tube is made of a ½ inch diameter and 90 cm long copper tube which has a selective coating to provide higher absorptivity with water as a working fluid. A second surface that is concentric to the receiver tube covers it externally and consists of two concentric glass tubes of different diameters. To avoid convective losses and take advantage of the greenhouse effect, the PTC has a glass cover 85 cm long and 57 cm wide with an aluminum frame.

The PTC was instrumented with type T thermocouples duly calibrated for the measurement of temperature in the regions of interest: at the input of the tube receiver (T_{input}), the environment (T_{env}), the glass surface (T_{glass}), and the reflective surface ($T_{reflect}$). Likewise, the temperature at the outlet of the receiver tube (T_{output}) was registered, which was the output variable or target in the ANN model. An Agilent 34970A card and Agilent Benchlink Data Logger software were used to acquire and record experimental data.

The feeding and storage tanks were of 750 and 450 L, respectively. They were interconnected to enable recirculation and thus raise the temperature of the feed tank when hot water was unused. Valves and a pressurizing motor pump model AQC-15 with a power of 2/15 HP were installed to control the flow. A Blue-White F-400 flowmeter with a capacity of 20 LPM was used to measure the volumetric flow at the outlet of the tube. The PTC was located on the facilities of the CIICAp-UAEM with latitude and longitude coordinates of 18.981655−99.23418, respectively. The PTC had an inclination of 19 degrees and was oriented east–west with the face of the reflecting plate pointing toward the south to better exploit the incidence of solar radiation. Fig. 2.4 shows the experimental equipment in operation.

A total of 67,691 samples were recorded during May and June from 8:00 to 19:00 hours, with intervals of 10 seconds between measurements. Table 2.1 shows the operating ranges to which the PTC system was subjected.

2.3 Development of the neural network model

The nervous system, the most complex system in the human body, comprises many billions of interconnected cells called neurons organized in anatomical pathways that are involved in sensory, motor, and cognitive functions. It is the pattern of connections between these neurons through

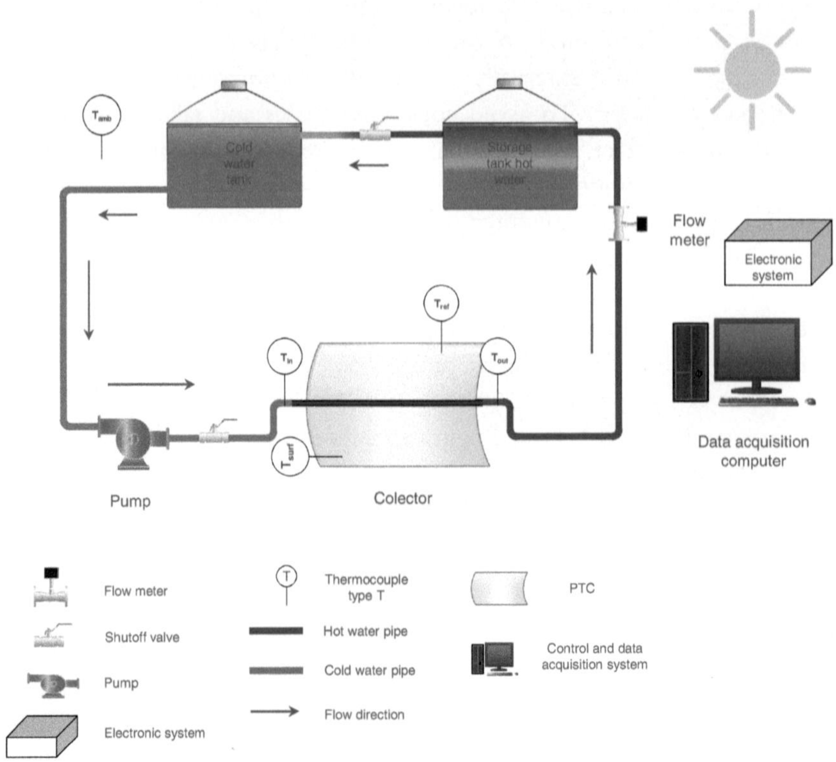

Figure 2.4 Schematic diagram of the experimental system PTC with low-cost components.

Table 2.1 Interval of experimental operating conditions.

Experimental variables	Working range
Day	1–31
Month	1–12
Hour (s)	20267–69085
T_{input} (°C)	15.43–50.00
T_{env} (°C)	9.94–70.28
T_{glass} (°C)	14.5–51.4
$T_{reflect}$ (°C)	16.00–77.92
T_{output} (°C)	15.59–50.66

a complex chemical process that determines brain functions [35]. The anatomical organization in a sequence of neurons and their ability to respond to external stimulation has inspired researchers to create models that reproduce this biological behavior.

The background related to the structure and physiology of the human learning system has helped as the basis for proposing expert systems in knowledge incorporated in themselves, known as connectionist systems. In this way, research in the neurobiological field has progressed in close collaboration with the mathematical and computer community. The central principle of connectionism is that the phenomena can be described by blocks of simple and frequently equal units that interconnect, generating what we today describe as models of ANNs [36].

Historically, the use of neural network models began with Warren McCulloch and Walter Pitts [37] in 1943, who proposed a computer element model to represent the activation of a neuron as a threshold function. Like its biological counterpart, the neuron performed a weighted sum of the inputs followed by a logical threshold operation. However, the main drawback of this calculation model was that the neuron could not learn from the examples since it established fixed values for the connections.

A few years later, in 1949, Donald Hebb showed that perception or concept is represented in the brain by a set of simultaneously active neurons [38]. He was the first to explain the learning processes, an essential element of human intelligence, from the psychological point of view. This discovery allowed the development of a procedure that would show how neurons strengthened connections through an iterative learning scheme, which adjusted the connection values. This scheme would become a fundamental rule of learning in the literature on neural networks [39].

In 1957, the first neural network model, known as perceptron, was published, which was developed by Frank Rosenblatt and inspired by Donald Hebb's studies. Perceptrons were the generic name assigned to a family of theoretical and experimental ANN models that Rosenblatt proposed in the period 1957–62 [40]. The perceptron is a classifying system capable of learning something from the recognition of geometric and abstract patterns. As the first simple learning algorithm, the perceptron marked the beginning of the use of neuronal schemes to successfully manage linearly separable classification problems [41]. Currently, the names of multilayer perceptron and perceptron are used to refer to specific ANN structures based on the Rosenblatt perceptrons.

Nowadays, ANNs are framed within the field of artificial intelligence (AI). AI is defined as the ability to mimic human intelligence in the movements of understanding, perception, problem solving, and decision-making to reproduce them through artificial neurons and relationships in an objective; having as a fundamental characteristic the ability of self-learning [42]. With advances in AI and computer science, ANNs can learn to solve complex problems in a reasonable amount of time.

ANNs have been defined as a mathematical modeling tool for nonlinear statistical data that can solve complex problems in different fields of knowledge [43]. ANNs can learn powerfully complex nonlinear and multivariable relationships from a set of training examples associated with phenomena that are challenging to solve through conventional methods [44].

They are composed of a large number of simple processing elements or nodes that are highly interconnected at levels operating in parallel [45] inspired by the shape, connection, and information processing of biological neurons. These elements process and send information to each other by modifying their dynamic responses to external variables through accumulated experience. The parallel architecture allows them to handle large and often incomplete volumes of data efficiently. They are capable of solving problems that do not have a clearly defined algorithm to transform input variables into an output response. Therefore, by combining adjustment parameters in an automatic process, ANNs learn, recognize, and apply relationships between variables.

ANNs are an excellent tool for the treatment of information, standing out for three important principles: adaptive learning, generalization, and abstraction [46].

1. Adaptive learning

   The network experience is obtained from adaptive learning. ANNs learn and train themselves by adapting their behavior according to the environment to produce a consistent output value.
2. Generalization

   Networks offer a response to data that you have not learned before, but that can be inferred based on your learning during training. Also, it can recognize information patterns with noise, distortion, or incomplete.
3. Abstraction

ANNs are able to abstract the main characteristics of a set of data that apparently do not present common or relative aspects.

Due to the inspiration in the fundamental cell of the nervous system, the processing elements are called artificial neurons. They are organized and interconnected in a similar way to how they occur in the biological brain, where each neuron receives and combines signals from and to other neurons undergoing various operations in order to reach a certain threshold and produce specific output response values [47]. Through a schematic representation, the analogy between a biological neuron and an artificial neuron is shown in Fig. 2.5.

The ANN is not an isolated element but is composed of several artificial neurons in interrelation. Generally, in an ANN, neurons are organized into

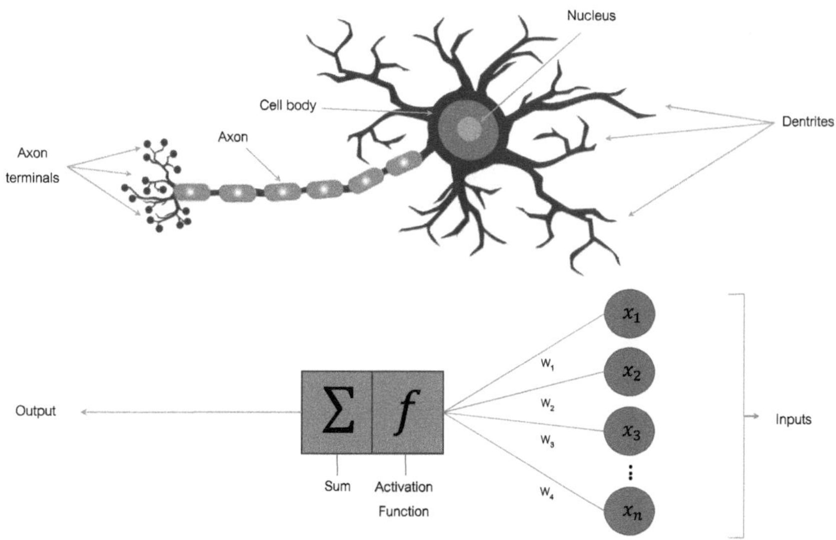

Figure 2.5 The analogy between a biological neuron and an artificial neuron.

three main groups called levels or layers [48]: the input layer, the hidden layer, and the output layer. Two layers establish connections with the outside:

- The input layer: where the input data to the network model are presented. These data are a finite value of the direct representation referring to the measurements of a process, system, or phenomenon delimited by the operating interval.
- The output layer: where the network response is presented concerning the input information.

Between the input and output layers, there are one or several layers called hidden layers which generally [49] constitute the majority of the neural network. A general representation of an ANN is shown in Fig. 2.6. Each circular node represents an artificial neuron, and each arrow represents a connection from the exit of each neuron to the entrance of others located in the different layers.

The interest of the ANN resides mainly in interneuronal connections and in the transmission of information that is carried between them. As is known biologically, when the neuron is stimulated through its dendrites, there is a change in the electrical potential. Sometimes the potential is not enough to take the neuron to the threshold, but it can be added together with others to trigger an action potential and transiently transform into a signal. Then, the receiving neuron adds or integrates all the inhibitory and excitatory signals it receives to decide whether or not to activate a signal toward the axon called action potential. This signal is transmitted to a receiving neuron allowing the transmission of unidirectional information packets, a process known as a synapse [50]. This dynamic signaling behavior represents a key factor in the efficient processing of information.

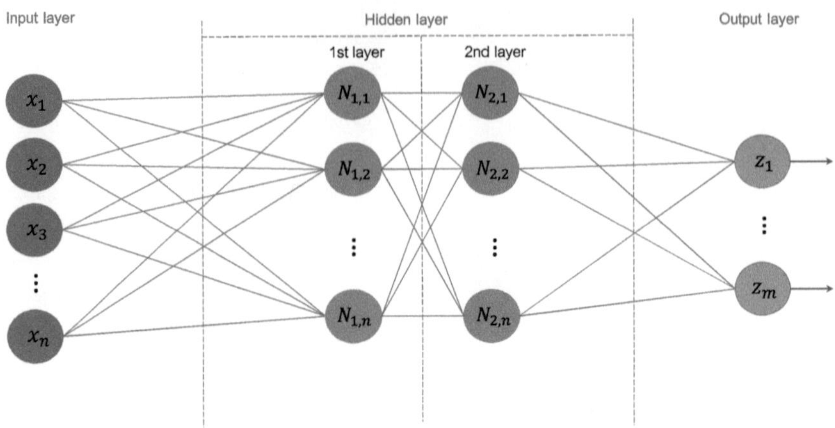

Figure 2.6 Representation of the artificial neural networks.

In the case of the artificial neuron, the analog to the electric potential is represented by a numerical value called synaptic weight, which can be positive if the neuron increases the activation state of the adjacent or negative if it inhibits it. Therefore, the weight is defined as an adjustable parameter that establishes the influence that one neuron has on another; representing the intensity of the signal. Within the network, each artificial neuron performs the mathematical calculation of the weighted sum of the input values with the synaptic weights to produce an output value simulating the action potential. In this way, the connection between neurons and layers has an associated weight value and a compensation value known as bias, which allows the ANN to learn [51]. Then, the input value for a neuron in the hidden layer is assigned by the following expression:

$$n_J = Wi_{(J,1)} \bullet P_{(1)} + Wi_{(J,2)} \bullet P_{(2)} + \ldots + Wi_{(J,R)} \bullet P_{(R)} + b_{(1,J)} \quad (2.1)$$

where Wi is the coefficients of the connection weights between the input layer and the hidden layer, P is the input variables, J is the number of neurons in the hidden layer, R is the number of input neurons, and $b_{(1,J)}$ is the bias corresponding to the neuron in the hidden layer.

In artificial synapses, the neuron calculates its degree of connectivity by the activation function or transfer function, which represents the potential rate at which a neuron is activated or not. Just as the axon transmits certain information to other biological neurons, the transfer function emits an output signal in a given range of amplitude; this will depend on the function used [46]. Later, the emitted signal will be the input to the next neuron of the subsequent layer of the neural network.

There are different types of transfer or activation functions such as step, linear, logarithmic sigmoidal, sigmoidal, Gaussian, and Gaussian inverse functions. The most prominent are for the hidden layer, the tangential-sigmoidal function (TANSIG, τ), and the logarithmic-sigmoidal (LOGSIG, δ). Normally, a linear transfer function (PURELIN, λ) is used for the output layer [52].

As mentioned earlier, the ANN can be trained with data that contain some noise; however, data preprocessing may be necessary. Good practice has been to eliminate deviant data (outliers) through statistical techniques [53]. Besides, if the application or system allows it, we can reduce the number of decimals of the experimentally acquired values.

It is essential to use normalization (or standardization) of the data to adjust at the same scale; that allows comparing their proportions instead of their magnitudes. Also, the normalization of the database prevents mathematical saturation as sigmoid transfer functions tend to become saturated when they have values higher than +3 or less than −3, as can be seen in Fig. 2.7. Therefore, having a high number of values on the input variables, the amount assigned to the weights must be minimal and, therefore, this would make the training very slow [54]. The normalization interval of the input variables must be related to the activation function that is decided to be used. Usually, it is done in intervals from −1 to 1, for which there are different equations in the literature.

After preprocessing data, learning of a neural network is carried out by modifying, and if necessary, the creations of new interneuronal connections are applied. Therefore, the network learns examining data patterns and generating the prediction by adjusting values assigned to weights and bias [55]. This learning process is achieved with the backpropagation algorithm.

The backpropagation algorithm consists of introducing the input variables to the network, obtaining a final output in the last layer, and comparing this result with the expected outcome (supervised learning) [56]. Then, the values of the weights and bias in the layers (output, hidden, and input) are modified backward to minimize the error between the desired output and the output generated (gradient descent method) [57]. This process is repeated consecutively until the network reaches one or more stop criteria [58] such as: reducing the error between the desired and the actual value to reach a defined number of iterations or obtain higher accuracy.

The proportion of each contribution will depend on the results of the adjustment of weights and biases obtained by the network to reach an output value with high precision. It is important to mention that, during training, overlearning should be avoided [59]. This is due to the network not being able to generalize problems with new data and then indicating erroneous results. A common practice to prevent overlearning has been to keep a low number of neurons in the hidden layer depending on the errors obtained during the learning process. Also, it should be considered that each neuron in the hidden layer significantly increases the number of coefficients in the output model.

Currently, research into ANNs focuses on improvements concerning the learning of the model. To ensure that the ANN learns efficiently, it

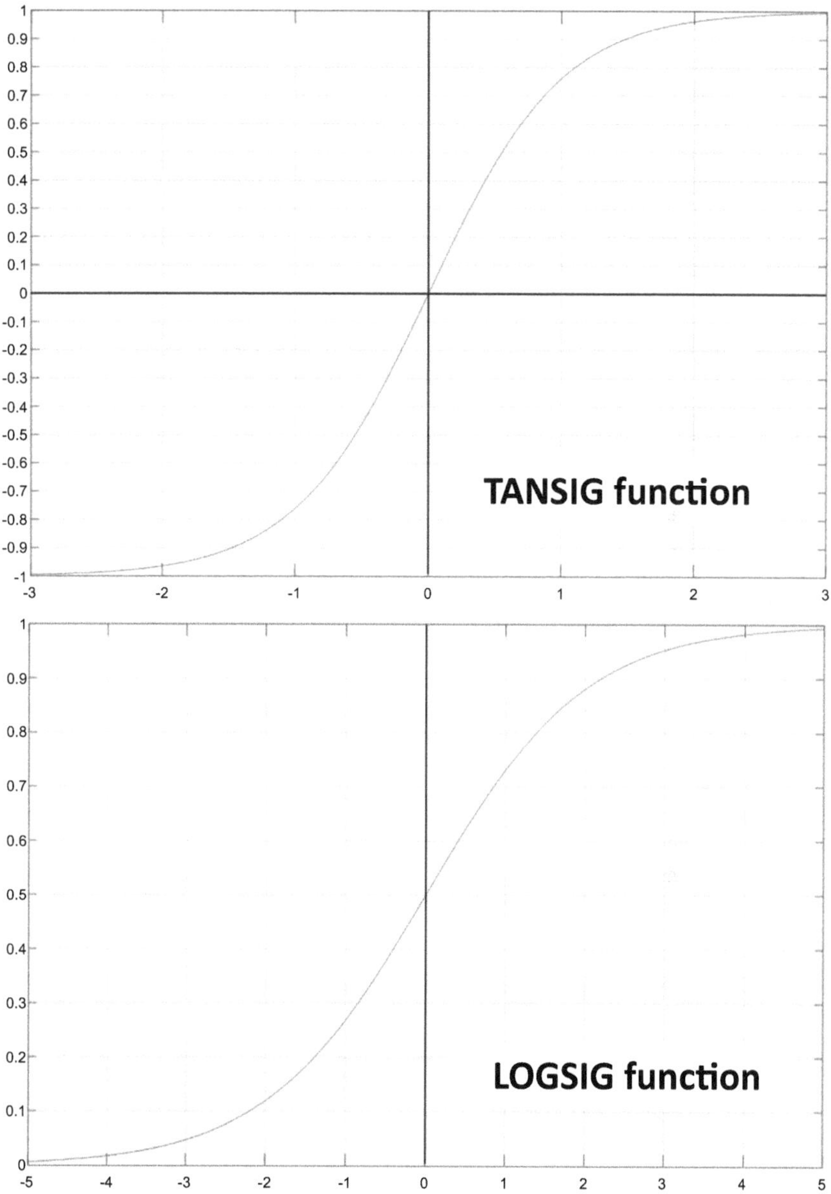

Figure 2.7 The behavior of data using TANSIG and LOGSIG functions.

has been necessary for the development of algorithmic training techniques to provide the network with the ability to discriminate and generalize the data. There are different classes of ANNs: single-layer feedforward networks, multilayer networks, and among others [60]. Some of the training or optimization algorithms that can be considered are conjugated gradient, quasi-Newton, and Levenberg–Marquardt [61]. The most commonly used ANN architecture is the multilayer perceptron trained with the backpropagation algorithm in a feedforward network through Levenberg–Marquardt learning (LM) [62].

This Levenberg–Marquardt learning method is similar to quasi-Newton methods, which were designed to approximate the speed of second-order training without having to calculate the Hessian matrix. When the performance function is in the form of a sum of squares (as is typical in feedforward training networks), then Hessian's matrix can be approximated as:

$$\boldsymbol{A} = \boldsymbol{H} = \boldsymbol{J}^T\boldsymbol{J} \tag{2.2}$$

where the gradient can be calculated as:

$$\boldsymbol{g} = \boldsymbol{J}^T\boldsymbol{e} \tag{2.3}$$

where J is the Jacobian matrix that contains the first derivatives of network errors with respect to weights and bias, and e is a vector of network errors. The Jacobian matrix can be calculated through a standard backpropagation technique [63] that is much less complex than the calculation of the Hessian matrix. The modification of the Levenberg–Marquardt method to approximate the Hessian matrix in the Newton method is:

$$x_{k+1} = x_k - [\boldsymbol{J}^T\boldsymbol{J} + \mu\boldsymbol{I}]^{-1}\boldsymbol{J}^T\boldsymbol{e} \tag{2.4}$$

When the μ scalar is zero, it is similar to Newton's method with the consideration that with the Hessian approximation, it is faster to calculate. When μ is large, the equation becomes a gradient descent method with a small step size. Newton's method is quicker and more accurate near a minimum error, so the goal is to move toward Newton's method as quickly as possible [64]. Its main disadvantage is that it requires storing Jacobian matrices that, for specific data sets, can be very large, which means a great use of memory [65].

To validate the accuracy of the model, statistical analysis is performed comparing experimental data versus simulated data. The test parameters that are commonly used are the root mean square error (RMSE), mean absolute percent error (MAPE), and the coefficient of determination

(R^2)[66]. The optimal model is the one with the highest R^2, and the lowest MAPE, and RMSE.

$$R^2 = 1 - \frac{\sum_{i=1}^{n} \left(x_{\exp(i)} - x_{\mathrm{sim}(i)}\right)^2}{\sum_{i=1}^{n} \left(x_{\exp(i)} - \overline{x}_{\exp}\right)^2} \tag{2.5}$$

$$\mathrm{MAPE} = \frac{\sum_{i=1}^{n} \left| \frac{x_{\exp(i)} - x_{\mathrm{sim}(i)}}{x_{\exp(i)}} \right|}{n} \times 100 \quad (\%) \tag{2.6}$$

$$\mathrm{RMSE} = \sqrt{\frac{\sum_{i=1}^{n} \left(x_{\mathrm{sim}(i)} - x_{\exp(i)}\right)^2}{n}} \tag{2.7}$$

To corroborate the fit of the model, significance tests are applied to the relation of experimental and simulated databases. The *F*-ratio test or Fisher test and Student's *t*-test are the significance tests most used to analyze the null hypothesis (H_0) or the alternative hypothesis (H_1)[67], where the critical value computed must be less than the critical value from the statistical tables to accept H_0. To calculate the *F* test and *t*-test, the following equations are used:

$$S = \sqrt{\frac{\sum_{i=1}^{n_x} (x_i - \overline{x}_i)^2 + \sum_{i=1}^{n_y} (y_i - \overline{y}_i)^2}{(n_x + n_y - 2)}} \tag{2.8}$$

$$F = \frac{S_{\exp}^2}{S_{\mathrm{sim}}^2} \tag{2.9}$$

$$t = \frac{\left| \overline{x_{\exp}} - \overline{x_{\mathrm{sim}}} \right|}{S\left(\sqrt{\left(\frac{1}{n_{\exp}} + \frac{1}{n_{\mathrm{sim}}}\right)}\right)} \tag{2.10}$$

where S is the standard deviation combined. For the *F*-test, the calculated F must always be ≥ 1. In the case that it is lower, it must be recalculated with the inverted terms. The numerator must always be larger than or equal to the denominator. When the *F*-test is passed, the student's *t*-test is carried out. Otherwise, it is no longer applied, and the model must be rejected. Finally, to validate the relationship between the simulated data with the ANN model and the experimental data, linearity tests are performed. The slope-intercept linearity test, also called the standard deviation of the linear regression or standard error, serves to establish the confidence limits of the slope and intercept [68].

After obtaining the ANN model, it is possible to determine the influence that the variables have on it through a study of the relative importance, called a sensitivity analysis. This equation is based on the neural network weight matrix proposed by Garson and mencioned in Parrales et al. [49]:

$$I_P = \frac{\sum_{h=1}^{H}\left(\left(\frac{|W_{(p,h)}|}{\sum_{r=1}^{R}|W_{(i,h)}|}\right)\times|W_{(h,o)}|\right)}{\sum_{i=1}^{I}\left\{\sum_{h=1}^{H}\left(\left(\frac{|W_{(i,h)}|}{\sum_{i=1}^{I}|W_{(i,h)}|}\right)\times|W_{(h,o)}|\right)\right\}} \tag{2.11}$$

where I_P is the relative importance of the p input variable on the output variable, and the subscripts i, h, and o represent the numbers of neurons in the input, hidden, and output layers, respectively.

2.4 Neural network model

To develop the ANN model required three fundamental tasks. The first task was the formation of a working database provided by the variation of operation variables and acquired through instrumentation equipment. It is essential to mention that the variables considered for the model were representative of the phenomenon to reduce the size of the network and avoid redundancy in training data. In this point, the normalization of the data set must be included. The second task was to obtain an adequate network configuration to reach an accurate prediction. In this step, the number of hidden neurons, the training algorithm, and the activation function as essential parameters for neural network training were considered. Therefore, later, in the third task, the correlation between experimental and predicted outlet temperature of the solar collector was established.

For the formation of a working database, 67,691 experimental tests during May and June 2014 that fed the ANN model were obtained. Among all the measurements made, seven variables were selected that were those with the most significant impacts for a solar collector model. The input variables were: the inlet temperature of the fluid, month, day, hour, ambient temperature, glass surface temperature, and reflective surface temperature.

The input data were normalized in a range of 0–1 due to the function used in the hidden layer being a hyperbolic tangent sigmoid transfer function TANSIG. Normalized data of [0,1] have shown better results compared to other normalizations [69]. It is known that the normalization represents a direct influence on the model and offers benefits such as suitable results and a considerable decrease in the calculation time [70]. After the normalization process, the dataset was randomly divided into training (50%), test (25%), and validation (25%). The equation that was used for data normalization [61] is the following:

$$x_{i,\text{norm}} = 0.8\left(\frac{x_i - x_{\min}}{x_{\max} - x_{\min}}\right) + 0.1 \tag{2.12}$$

To obtain an adequate network configuration and achieve the right prediction of the solar collector outlet temperature as reliably as possible, different ANN architectures must be developed. For this, a feedforward network model was used with implementation of the backpropagation structure. The hyperbolic tangent sigmoid transfer function and linear transfer function were used in the hidden layer and output layer, respectively. As mentioned earlier, this algorithm is widely used for having a shorter computational processing time. By its design of approximation of the second partial derivatives, the LMA is not used to calculate the Hessian matrix [71]. In addition, the Levenberg–Marquardt backpropagation algorithm (LMA) has a smaller mean square error (MSE) compared to other backpropagation algorithms [72]. The development of the PTC neural model was carried out using the MATLAB® Software (R2015b, Mathworks, Natick, MA, USA) and the ANN Toolbox Toolkit. In Fig. 2.8, the learning process of the model ANN is schematized and Fig. 2.9 shows a schematic of the model ANN.

For the last task, the network performance was evaluated using the statistical parameters as the coefficient of determination (R^2), the MAPE, and the RMSE. These parameters were described in Section 2.3. The best architecture was presented with 19 neurons in the hidden layer, and in Table 2.2 the comparison of the statistical analysis is shown. As can be seen, when increasing the number of hidden neurons above 20, the values

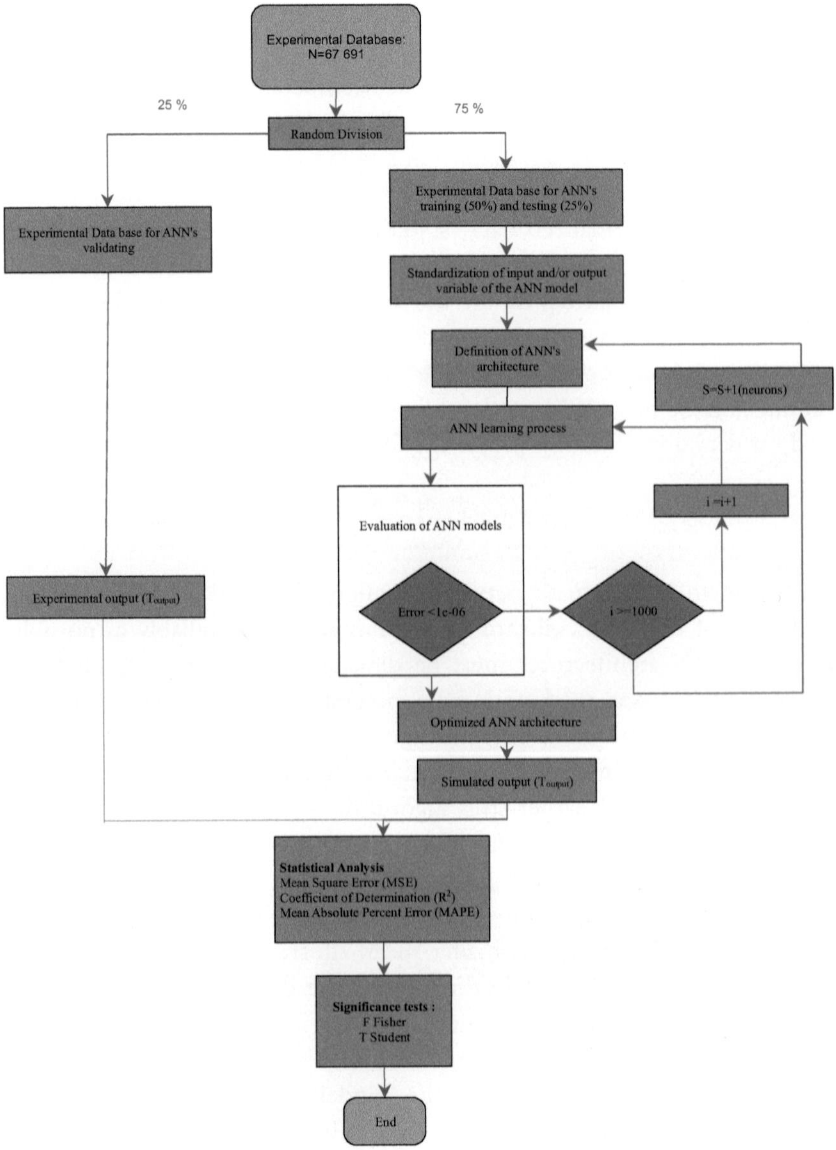

Figure 2.8 Methodology for training the ANN model of a parabolic trough solar collector system.

obtained from R^2, RMSE, and MAPE do not show significant variations. Therefore, the continued increase in the number of neurons in the hidden layer would reduce the error by very little.

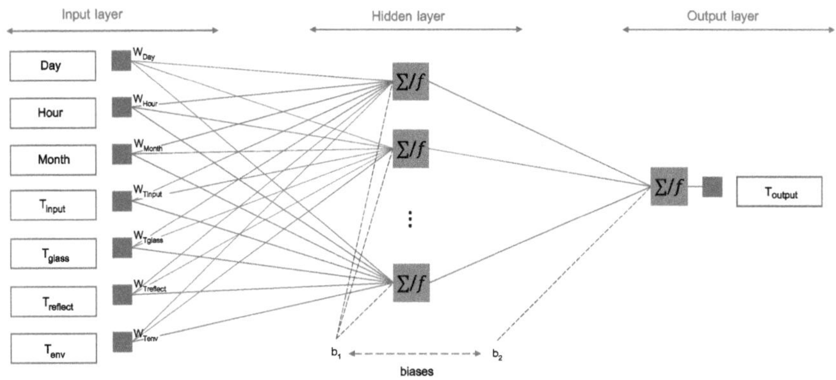

Figure 2.9 Schematic of the ANN model of a parabolic trough solar collector system.

Table 2.2 Comparison of the statistical analyses for determining the number of hidden neurons.

Architecture ANN	Number of neurons	Coefficient of determination	Root mean square error	Mean absolute percent error
		R^2	RMSE	MAPE
7–1–1	1	0.8833	1.9824	1.2122
7–2–1	2	0.9092	1.8182	1.1209
7–3–1	3	0.9407	2.0009	1.0951
7–4–1	4	0.9460	1.9167	1.0526
7–5–1	5	0.9577	1.7930	1.0065
7–6–1	6	0.9654	1.7793	0.9871
7–7–1	7	0.9664	1.8196	0.9832
7–8–1	8	0.9704	1.8501	0.9729
7–9–1	9	0.9725	1.8345	0.9888
7–10–1	10	0.9773	1.7538	0.9563
7–11–1	11	0.9972	1.7835	0.9592
7–15–1	15	0.9865	0.7700	0.0555
7–16–1	16	0.9865	0.7701	0.0603
7–17–1	17	0.9867	0.7649	0.0541
7–18–1	18	0.9883	0.7176	0.0542
7–19–1	**19**	**0.9894**	**0.6829**	**0.0405**
7–20–1	20	0.9886	0.7078	0.0462
7–25–1	25	0.9932	0.5452	0.0319
7–30–1	30	0.9915	0.6119	0.0353

In addition, considering a higher number of hidden neurons could involve a more significant amount of coefficients in the resulting equation, the best model was presented with a 7–19–1 architecture with a R^2

coefficient of determination of 0.9894. Importantly, the MAPE generalization error is minimal, indicating that the network can generalize from new data to obtain good predictions.

To improve the performance function, the Bayesian regularization training method (BR) that is a modification to the Levenberg–Marquardt algorithm, was used, which enhanced the generalization of the ANN [73]. This method consisted of the alteration of the performance index, given by:

$$F = \beta E_d + \alpha E_w \tag{2.13}$$

where α and β are parameters to be optimized, E_d is the sum of the squared errors of the network, and E_w is the sum of the squared weights. This algorithm provides a measure of which parameters (weights and biases) are being used effectively by the network.

The comparison between the statistical parameters of R^2, RMSE, and MAPE using both algorithm training is described in Table 2.3. With these results, it can be seen that the Bayesian regularization algorithm is able to minimize further the errors achieved with the Levenberg–Marquardt algorithm (MAPE = 0.0372). Although the error in RMSE is slightly higher (RMSE = 0.6829), the MAPE generalization error is minimal. Therefore, it indicates that the network could generalize from new data and obtain accurate predictions.

Fig. 2.10 shows the comparison of the experimental output temperature and that simulated by the ANN models with the Bayesian regularization training method.

Fig. 2.11 shows the simulation of the outlet temperature of the collector with the models proposed for different days. It is shown that, in general, ANN models follow a defined pattern of water heating at the outlet of the collector despite the variations that may exist. The variations are present because, in renewable energy systems, nature is exposed to

Table 2.3 Comparison between statistical parameters of R^2, RMSE, and MAPE using LM and BR algorithm training.

Architecture ANN	Number of neurons	Coefficient of determination	Root mean square error	Mean absolute percent error
		R^2	RMSE	MAPE
7–19–1	19	0.9894	0.6829	0.0405
7–19–1	19	0.9909	0.6285	0.0372

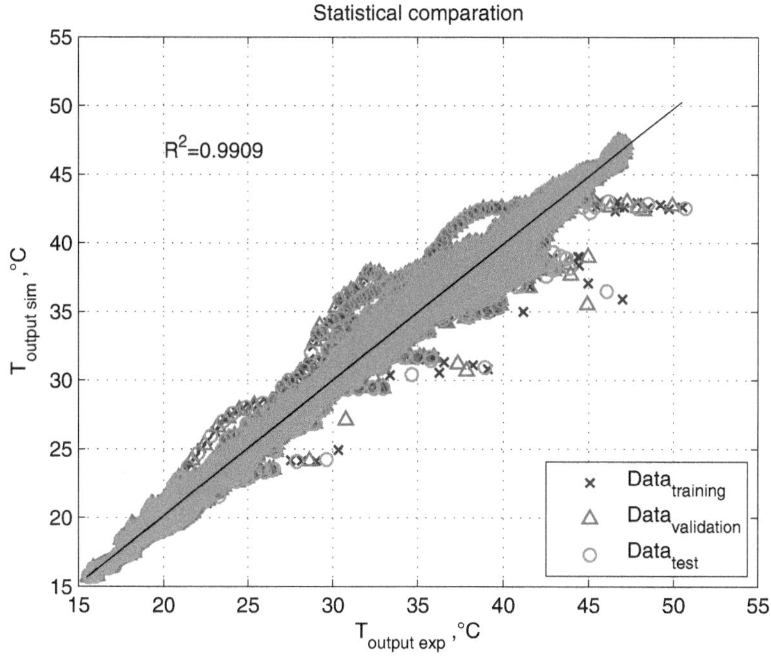

Figure 2.10 Statistical comparison between simulated (ANN) and experimental outlet temperature data with BR algorithm training.

different factors that change concerning time and space, causing noise in the measurements. Namely, they generate data far from the average of the samples.

Nevertheless, the predictions of the network can be improved through the Bayesian regularization algorithm. It is observed in detail that this algorithm follows better the main pattern of experimental data, and tries to discriminate the peaks that are generated compared to its counterpart. The simulated data with the Bayesian regularization algorithm are, for the most part, less dispersed and when there is a complicated pattern to follow it manages to remain within the average.

To verify the linearity and establish the confidence limits for the regression parameters, the linearity test was performed. In the linearity test, the slope value must be 1 (or close to 1), and the intersection must be 0 (or close to 0). Table 2.4 shows the linearity test for both training algorithms, and, as can be seen, Bayesian regularization obtained better results compared to the Levenberg–Marquardt algorithm, since its parameters are closer to 1 and 0, for slopes and intercepts, respectively.

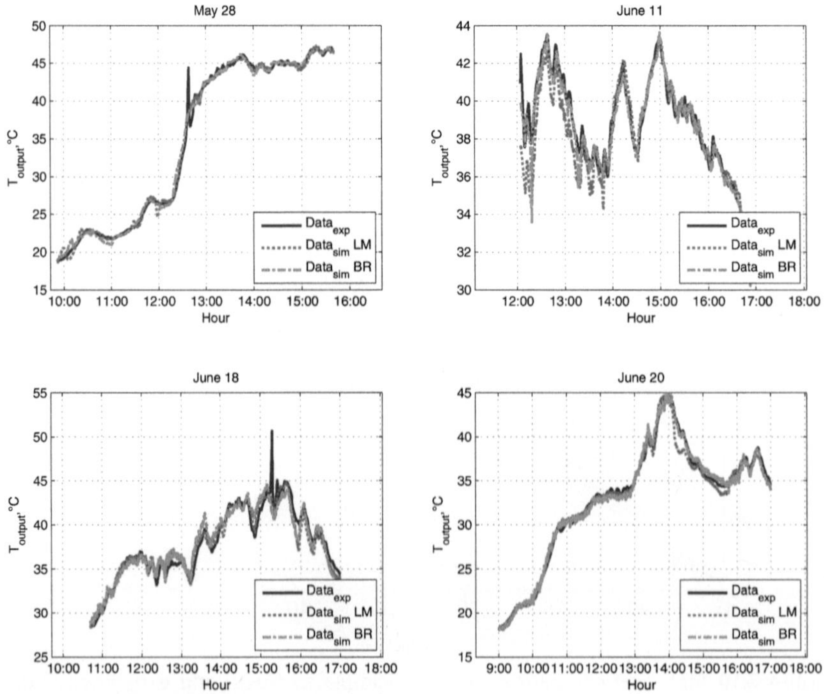

Figure 2.11 Simulation of the outlet temperature of the collector with the models proposed for different days.

Table 2.4 Slope and intercept statistical test.

Training method	R^2	99% Confidence			
		Slope upper	**Slope lower**	**Intercept upper**	**Intercept lower**
LM	0.9894	0.9904	0.9884	0.3529	0.2902
BR	0.9909	0.9914	0.9901	0.3013	0.2435

Therefore, the best model for predicting the temperature of the solar collector was the 7−19−1 architecture with the Bayesian regularization training algorithm. The ANN model is represented in general terms according to the activation functions selected, which were [74]:

$$\lambda = Wi_{(h,1)} \bullet P_{(1)} + Wi_{(h,2)} \bullet P_{(2)} + Wi_{(h,3)} \bullet P_{(3)} + \ldots + Wi_{(h,i)} \bullet P_{(i)} + b_{(1,h)} \tag{2.14a}$$

$$\lambda = \sum_{i=1}^{I} Wi_{(h,i)} \bullet P_{(i)} + b_{(1,h)} \tag{2.14b}$$

$$\text{tansig}(\tau) = \frac{1 - \exp(-\lambda)}{1 + \exp(-\lambda)} = \frac{2}{1 + e^{-2\lambda}} - 1 \tag{2.15a}$$

$$\text{tansig}(\tau) = \frac{1 - \exp\left[-\left(\sum_{i=1}^{I} Wi_{(h,i)} \bullet P_{(i)} + b_{(1,h)}\right)\right]}{1 + \exp\left[-\left(\sum_{i=1}^{I} Wi_{(h,i)} \bullet P_{(i)} + b_{(1,h)}\right)\right]} = \frac{2}{1 + e^{-2\left(\sum_{i=1}^{I} Wi_{(h,i)} \bullet P_{(i)} + b_{(1,h)}\right)}} - 1 \tag{2.15b}$$

Therefore, the fluid outlet temperature (T_{output}) was obtained by replacing the 172 coefficients obtained (weights and bias) in the following equation:

$$T_{\text{output}} = \sum_{h=1}^{H} \left[Wo_{(o,h)} \left(\frac{2}{1 + \exp\left(-2\left(\sum_{i=1}^{I} \left(Wi_{(h,i)} \bullet P_{(i)}\right) + b_{1,(h)}\right)\right)} - 1 \right) \right] + b_{2,(o)} \tag{2.16}$$

where Wo is the weights between the hidden layer and the output layer, Wi is the weights between the input-hidden layers, b_1 is the bias in the hidden layer and b_2 is the bias in the output layer. The subscripts i, h, and o represent the number of neurons in the input, hidden, and output layers, respectively. Formulating the equation is not a complicated process as only simple arithmetic operations are required. Table 2.5 shows the coefficients resulting from the Bayesian regularization training.

With the Garson equation it was possible to discern the influence of variables on the model ANN PTC. The final percentages of input variables are shown in Fig. 2.12.

Through the result of the sensitivity analysis, it was determined that the variable with greater preponderance over the model was the ambient temperature with 26.05%. This was expected, since the ambient temperature obtained throughout the day is closely related to solar radiation. The temperatures at around 12:00 hours are maximum, and the minimums were around 19:00 hours.

Also, it was shown that both the day and month are of importance, with 23.47% and 15.32%, respectively; this reiterates the dependence of the model on the season of the year and these can be considered as influential variables. The inlet temperature of the fluid (12.02%), the reflective

Table 2.5 Parameters of weights and bias obtained with BR algorithm training.

Neurons	Weights								Bias	
	W_i							W_o		
	Day	Month	Hour (s)	T_{input} (°C)	T_{env} (°C)	T_{glass} (°C)	$T_{reflect}$ (°C)	T_{output} (°C)	b_1	b_2
1	−6.6343	6.2872	3.4883	−2.2966	0.0203	−1.5021	3.8636	−12.7682	−6.0673	2.3190
2	0.9612	−2.0988	−4.4438	15.1499	16.4538	−0.0580	−6.4706	40.2968	−2.7195	
3	42.6016	26.0394	78.3668	−48.4249	−14.4120	36.1544	11.4646	0.5659	−73.3716	
4	−5.8659	0.9408	3.9892	−10.6760	−17.9960	2.4015	2.7706	3.6064	9.0308	
5	−29.4409	−6.4120	−5.3869	5.7846	13.8036	−2.5486	−0.4248	−4.9614	18.4303	
6	9.2342	5.7294	0.7812	1.7514	1.8980	−3.6860	−2.1114	−8.1805	−9.4105	
7	8.6283	−6.2575	6.5926	−6.6689	12.5297	2.5046	2.5926	2.2803	−1.6101	
8	−6.9416	−12.3673	2.6605	−5.0401	−7.9831	−1.2154	3.6659	−11.1146	7.5469	
9	−2.1297	−1.5680	0.0114	2.9297	7.6631	−0.4124	3.2281	10.8348	−0.4840	
10	2.2005	0.1645	−0.7058	−0.1350	−0.7118	0.7017	−1.4186	−45.4173	0.2979	
11	−5.3189	4.0159	1.4266	−0.5928	4.6552	−0.4311	1.1626	−54.5733	−2.8135	
12	6.5120	−1.6923	−1.2098	−0.5772	−6.8965	0.9929	−0.3655	−40.2758	0.4191	
13	−209.2692	30.1575	−44.5119	33.9495	155.4319	70.3200	−123.0939	−2.6301	24.9688	
14	−10.8554	−26.1708	−64.2758	28.8494	−60.9620	−11.5636	−5.2275	−1.0686	79.0484	
15	9.9964	0.6544	−13.9439	24.0715	47.3801	39.6620	−37.4032	1.4398	−17.3910	
16	6.6736	11.0591	−1.4247	−0.0704	−6.0150	1.2079	−1.1522	47.0567	−1.1056	
17	−0.8569	1.8382	4.0467	−13.8967	−16.0811	0.0087	5.5881	46.2467	2.9249	
18	−191.9632	−70.4659	−0.8560	−0.0997	−12.3246	0.1633	0.0273	−6.8981	185.159	
19	10.2503	0.7817	4.6163	−9.3261	0.9948	0.5215	3.5090	−4.5800	−8.1016	

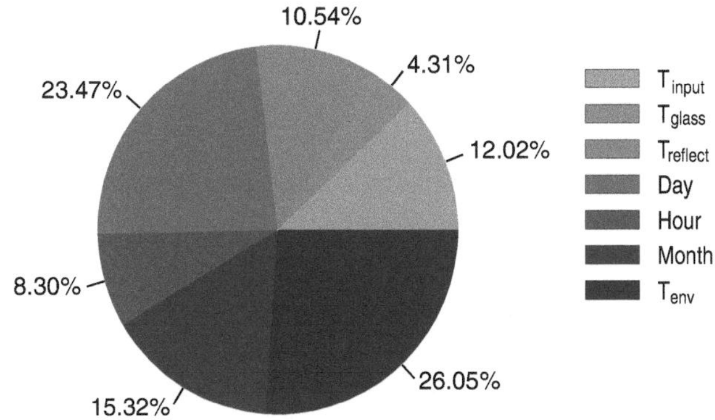

Figure 2.12 The relative importance of the input variables in the ANN model of PTC system.

surface temperature (10.54%), and the time (8.30%) variables showed a small impact with similar proportion. Finally, the least important variable was the surface temperature of the glass, with 4.31%.

2.5 Conclusions

An ANN model was developed to predict the water outlet temperature in a low-cost solar collector by experimental measurements of a water heating process. The model considers seven input variables, which were: the inlet temperature of the fluid, month, day, hour, ambient temperature, glass surface temperature, and reflective surface temperature.

The neural network model was trained with the Levenberg–Marquardt algorithm. The best architecture found was with 19 neurons in the hidden layer. According to the statistical tests, the following results were obtained: $R^2 = 0.9894$, RMSE = 0.6829, and MAPE = 0.0405%. The network obtained using the Bayesian regularization training algorithm was improved, obtaining the following statistical results: $R^2 = 0.9909$, RMSE = 0.6285, and MAPE = 0.0372%. The results obtained indicate that the obtained ANN model was able to simulate the nonlinear behaviors of the collector outlet temperature for the May to June period.

With the sensitivity analysis, it was shown that all variables influence the model. However, the ambient temperature was the variable that had most influence, followed by the variables are related to the season. The water inlet temperature and reflective surface temperature depended on the materials and design of the PTC, so these variables can be considered as possible opportunities to increase the outlet temperature.

When applying an ANN model to predict the outlet temperature of the collector, it is not intended to eliminate the use of mathematical models to obtain the performance of a solar collector but rather to be considered as an option that facilitates understanding of the behavior of the heating fluid in solar energy collectors.

References

[1] L. Kumar, M. Hasanuzzaman, N.A. Rahim, Global advancement of solar thermal energy technologies for industrial process heat and its future prospects: a review, Energy Conversion and Management 195 (2019) 885–908.
[2] S. Adams, C. Nsiah, Reducing carbon dioxide emissions; does renewable energy matter? Science of the Total Environment 693 (2019) 133288.
[3] Global Energy & CO2 Status Report. The latest trends in energy and emissions in 2018. https://www.iea.org/geco/emissions/.
[4] Sustainable Development Goals, 2019. https://www.un.org/sustainabledevelopment/climate-change/.
[5] IRENA, Climate Change and Renewable Energy: National policies and the role of communities, cities and regions (Report to the G20 Climate Sustainability Working Group (CSWG)). International Renewable Energy Agency, Abu Dhabi, 2019.
[6] A. Wahab, A. Hassan, M.A. Qasim, H.M. Ali, H. Babar, M.U. Sajid, Solar energy systems – potential of nanofluids, Journal of Molecular Liquids 289 (2019) 111049.
[7] A. Sharif, S.A. Raza, I. Ozturk, S. Afshan, The dynamic relationship of renewable and nonrenewable energy consumption with carbon emission: a global study with the application of heterogeneous panel estimations, Renewable Energy 133 (2019) 685–691.
[8] IRENA, Renewable capacity highlights, 2019.
[9] P. Moriarty, D. Honnery, 6 – Global renewable energy resources and use in 2050, Academic Press Managing Global Warming (2019) 221–235.
[10] NASA Science Solar System Exploration, https://solarsystem.nasa.gov/solar-system/sun/in-depth/.
[11] R. Foster, M. Ghassemi, A. Cota, Solar Energy: Renewable Energy and the Environment, CRC Press, Taylor & Francis Group, 2010.
[12] H.A. Rypkema, Chapter 2.1—environmental chemistry, renewable energy, and global policy, Green Chemistry (2018) 19–47.
[13] B. Parida, S. Iniyan, R. Goic, A review of solar photovoltaic technologies, Renewable and Sustainable Energy Reviews 15 (2011) 1625–1636.
[14] A. Ray, S. De, Distributed polygeneration using solar energy: a future sustainable energy system for India. In Applications of Solar Energy Springer, Singapore, 2018, pp. 11–28.
[15] A. Shafieian, M. Khiadani, A. Nosrati, Strategies to improve the thermal performance of heat pipe solar collectors in solar systems: a review, Energy Conversion and Management 183 (2019) 307–331.
[16] L. Evangelisti, R.L. Vollaro, F. Asdrubali, Lastest advances on solar thermal collectors: a comprehensive review, Renewable and Sustainable Energy Reviews 114 (2019) 109318.
[17] E. Bellos, C. Tzivanidis, Alternative designs of parabolic trough solar collectors, Progress in Energy and Combustion Science 71 (2019) 81–117.
[18] D. Real, R. Johnston, J. Lauer, A. Schicho, N. Hotz, Novel non-concentrating solar collector for intermediate-temperature energy capture, Solar Energy 108 (2014) 421–431.

[19] S.A. Kalogirou, Solar thermal collectors and applications, Progress in Energy and Combustion Science 30 (2004) 231–295.
[20] O.A. Jaramillo, M. Borunda, K.M. Velazquez-Lucho, M. Robles, Parabolic trough solar collector for low enthalpy processes: an analysis of the efficiency enhancement by using twisted tape inserts, Renewable Energy 93 (2016) 125–141.
[21] L. Xu, F. Sun, L. Ma, X. Li, D. Lei, G. Yuan, et al., Analysis of optical and thermal factors effects on the transient performance of parabolic trough solar collectors, Solar Energy 179 (2019) 195–209.
[22] A.Z. Hafez, A.M. Attia, H.S. Eltwab, A.O. ElKousy, A.A. Afifi, A.G. AbdElhamid, et al., Design analysis of solar parabolic trough thermal collectors, Renewable and Sustainable Energy Reviews 82 Part 1 (2018) 1215–1260.
[23] A. Sandá, S.L. Moya, L. Valenzuela, Modelling and simulation tools for direct steam generation in parabolic-trough solar collectors: a review, Renewable and Sustainable Energy Reviews 113 (2019) 109226.
[24] J.D. Osorio, A. Rivera-Alvarez, Performance analysis of parabolic trough collectors with double glass envelope, Renewable Energy 130 (2019) 1092–1107.
[25] S. Quezada-García, H. Sánchez-Mora, M.A. Polo-Labarrios, R.I. Cázares-Ramírez, Modeling and simulation to determine the thermal efficiency of a parabolic solar through collector system, Case Studies in Thermal Engineering (2019) 100523.
[26] A.H. Elsheikh, S.W. Sharshir, M.A. Elaziz, A.E. Kabeel, W. Guilan, Z. Haiou, Modeling of solar energy systems using artificial neural network: a comprehensive review, Solar Energy 180 (2019) 622–639.
[27] N. Fadaei, W. Yan, M.M. Tafarroj, A. Kasaeian, The application of artificial neural networks to predict the performance of solar chimney filled with phase change materials, Energy Conversion and Management 171 (2018) 1255–1262.
[28] S.A. Kalogirou, Artificial neural networks in renewable energy systems applications: a review, Renewable and Sustainable Energy Reviews 5 (4) (2001) 373–401.
[29] S.A. Kalogirou, Prediction of flat-plate collector performance parameters using artificial neural networks, Solar Energy 80 (3) (2006) 248–259.
[30] D.V. Siva Krishna Rao K, M. Premalatha, C. Naveen, Analysis of different combinations of meteorological parameters in predicting the horizontal global solar radiation with ANN approach: a case study, Renewable and Sustainable Energy Reviews 91 (2018) 248–258.
[31] F.J. Diez, L.M. Navas-Gracia, A. Martínez-Rodríguez, A. Correa-Guimares, L. Chico-Santamarta, Modelling of a flat-plate solar collector using artificial neural networks for different working fluid (water) flow rates, Solar Energy 188 (2019) 1320–1331.
[32] A. Sözen, T. Menlik, S. Ünvar, Determination of efficiency of flat-plate solar collectors using neural network approach, Expert Systems with Applications 35 (4) (2008) 1533–1539.
[33] C. Cetiner, F. Halici, H. Cacur, I. Taymaz, Generating hot water by solar energy and application of neural network, Applied Thermal Engineering 25 (8–9) (2005) 1337–1348.
[34] S.Y. Heng, Y. Asako, T. Suwa, K. Nagasaka, Transient thermal prediction methodology for parabolic trough solar collector tube using artificial neural network, Renewable Energy 131 (2019) 168–179.
[35] S. Fischer, P. Frey, H. Drück, A comparison between state-of-the-art and neural network modelling of solar collectors, Solar Energy 86 (2012) 3268–3277.
[36] L. Sugavaneswaran, Mathematical Modeling of Gene Networks, Elsevier Inc, 2017.
[37] W.S. McCulloch, W. Pitts, A logical calculus of the ideas immanent in nervous activity, The Bulletin of Mathematical Biophysics 5 (4) (1943) 115–133.
[38] D.L. Glanzman, Associative learning: hebbian flies, Current Biology 15 (11) (2005) R416–R419.

[39] B. Yegnanarayana, Artificial Neural Networks. Prentice-Hall of India Private Limited, 2006.
[40] L.N. Kanal, Perceptrons, International Encyclopedia of the Social & Behavioral Sciences (2001) 11218–11221.
[41] V.N. Gudivada, M.T. Irfan, E. Fathi, D.L. Rao, Chapter 5—cognitive analytics: going beyond big data analytics and machine learning, Handbook of Statistics 35 (2016) 169–205.
[42] J. Shabbir, T. Anwer, Artificial intelligence and its role in near future, Journal of Latex Class Files 14 (8) (2015) 1–11.
[43] J. Penm, B. Chaar, R. Moles, J. Penm, 37—Predicting ASX health care stock index movements after the recent financial crisis using patterned neural networks, Rethinking Valuation and Pricing Models (2013) 599–610.
[44] B.A. Jensen, J. Abonyi, Neural Networks for Process Modeling, Process Control and Optimization, Boca Ratón, 2005, pp. 253–264.
[45] E.S. Elmolla, M. Chaudhuri, M.M. Eltoukhy, The use of artificial neural network (ANN) for modeling of COD removal from antibiotic aqueous solution by the Fenton process, Journal of Hazardous Materials 179 (1–3) (2010) 127–134.
[46] K.L. Du, M.N.S. Swamy, Neural Networks and Statistical Learning, Springer, 2014.
[47] J. Glinski, J. Horabik, J. Lipiec, Basic Concepts of Neural Networks. Encyclopedia of Agrophysics, Springer, 2011.
[48] M.A. Ghorbani, R. Khatibi, B. Hosseini, M. Bilgili, Relative importance of parameters affecting wind speed prediction using artificial neural networks, Theoretical and Applied Climatology 114 (1–2) (2013) 107–114.
[49] A. Parrales, D. Colorado, J. Díaz-Gómez, A. Huicochea, A. Álvarez, J.A. Hernández, New void fraction equations for two-phase flow in helical heat exchangers using artificial neural networks, Applied Thermal Engineering 130 (2018) 149–160.
[50] G. Nierhaus, Chapter 9 Artificial Neural Networks. Algorithmic Composition-Paradigms of Automated Music Generation, Springer Wien, NewYork, 2009, pp. 205–224.
[51] P. Khetarpal, M.M. Tripathi, A critical and comprehensive review on power quality disturbance detection and classification, Sustainable Computing: Informatics and Systems 28 (2020) 100417.
[52] S.J. Qin, Chapter 8—neural networks for Intelligent sensors and control—practical issues and some solutions, Neural Systems for Control (1997) 213–234.
[53] S.P. Verma, Estadística básica para el manejo de datos experimentales: aplicación en la Geoquímica (Geoquimiometría), Universidad Nacional Autónoma de México, México, 2005.
[54] J. Han, M. Kamber, J. Pei, 3—Data preprocessing. Data mining, The Morgan Kaufmann Series in Data Management Systems, third ed., 2012, pp. 83–124.
[55] R. Jafar, I. Shahrour, I. Juran, Application of artificial neural networks (ANN) to model the failure of urban water mains, Mathematical and Computer Modelling 51 (9–10) (2010) 1170–1180.
[56] S. Curcio, G. Iorio, 16—Models of membrane reactors based on artificial neural networks and hybrid approaches, Handbook of Membrane Reactors – Fundamental Materials Science, Design and Optimisation 1 (2013) 569–597.
[57] S. Jothilakshmi, V.N. Gudivada, Chapter 10—large scale data enabled evolution of spoken language research and applications, Handbook of Statistics 35 (2016) 301–340.
[58] K. Chakraborty, S. Bhattacharyya, R. Bag, A.A. Hassanien, 7—Sentiment analysis on a set of movie reviews using deep learning techniques, Social Network Analytics (2019) 127–147.

[59] M.H. Bakr, M.H. Negm, Chapter three—modeling and design of high-frequency structures using artificial neural networks and space mapping, Advances in Imaging and Electron Physics 174 (2012) 223–260.
[60] R.F. Turkson, F. Yan, M.K.A. Ali, J. Hu, Artificial neural network applications in the calibration of spark-ignition engines: an overview, Engineering Science and Technology, an International Journal 19 (3) (2016) 1346–1359.
[61] A. Sözen, M. Özalp, E. Arcaklioğlu, Calculation for the thermodynamic properties of an alternative refrigerant (R508b) using artificial neural network, Applied Thermal Engineering 27 (2–3) (2007) 551–559.
[62] G. Jozanikohan, G.H. Norouzi, F. Sahabi, H. Memarian, B. Moshiri, The application of multilayer perceptron neural network in volume of clay estimation: case study of Shurijeh gas reservoir, Northeastern Iran, Journal of Natural Gas Science and Engineering 22 (2015) 119–131.
[63] S. Mammadli, Financial time series prediction using artificial neural network based on Levenberg-Marquardt algorithm, Procedia Computer Science 120 (2017) 602–607.
[64] M.T. Hagan, H.B. Demuth, M.H. Beale, O. De Jesús, Neural Network Design, second ed., Martin Hagan, 2014.
[65] H. Demuth, M. Beale, Neural Network Toolbox for Use with MATLAB, 4, The MathWorks, Inc, 2005.
[66] A. Bassam, R.A. Conde-Gutierrez, J. Castillo, G. Laredo, J.A. Hernández, Direct neuronal network modelling for separation of linear and branched paraffins by adsorption process for gasoline octane number improvement, Fuel 124 (2014) 158–167.
[67] S.P. Verma, R. Cruz-Huicochea, Alternative approach for precise and accurate Student's t critical values and application in geosciences, Journal of Iberian Geology 39 (1) (2013) 31–56.
[68] C.I. Rocabruno-Valdés, J.G. González-Rodriguez, Y. Díaz-Blanco, A.U. Juantorena, J.A. Muñoz-Ledo, Y. El-Hamzaoui, et al., Corrosion rate prediction for metals in biodiesel using artificial neural networks, Renewable Energy 140 (2019) 592–601.
[69] J.A. Hernández, D. Colorado, Uncertainty analysis of COP prediction in a water purification system integrated into a heat transformer using several artificial neural networks, Desalination Water Treatment 51 (2013) 1443–1456.
[70] D.E. Millán-Ocampo, A. Parrales, J.G. González-Rodríguez, S. Silva-Martínez, J. Porcayo-Calderón, J.A. Hernández-Pérez, Modelling of behavior for inhibition corrosion of bronze using artificial neural network (ANN), Entropy 20 (2018) 409–419.
[71] J. Díaz-Gómez, A. Parrales, A. Álvarez, S. Silva-Martínez, D. Colorado, J.A. Hernández, Prediction of global solar radiation by artificial neural network based on a meteorological environmental data, Desalination and Water Treatment (2014) 1–8.
[72] M. Arabameri, A. Javid, A. Roudbari, The Use of Artificial Neural Network (ANN) for modeling of ammonia nitrogen removal from landfill leachate by the ultrasonic process, International Journal of Health Studies 1 (3) (2015) 13–19.
[73] D.V. Singh, G. Maheshwari, R. Shrivastav, D.K. Mishra, Neural network – comparing the performances of the training functions for prediction the value of specific heat of refrigerant in vapor absorption refrigeration system, International Journal of Computer Aplications 18 (4) (2011) 1–5.
[74] C.I. Rocabruno-Valdés, L.F. Ramírez-Verduzco, J.A. Hernández, Artificial neural network models to predict density, dynamic viscosity, and cetane number of biodiesel, Fuel 147 (2015) 9–17.

CHAPTER THREE

Applications of artificial neural networks in concentrating solar power systems

Mohamed E. Zayed[1,2], Jun Zhao[1], Wenjia Li[1], S. Sadek[1] and Ammar H. Elsheikh[3]

[1]Key Laboratory of Efficient Utilization of Low and Medium Grade Energy, MOE, Tianjin University, Tianjin, P.R. China
[2]Mechanical Power Engineering Department, Faculty of Engineering, Tanta University, Tanta, Egypt
[3]Production Engineering and Mechanical Design Department, Tanta University, Tanta, Egypt

Contents

3.1 Introduction

Currently, the attention on green energy systems as an alternative technology for nonrenewable energy sources has increased [1,2]. This is attributed to the advantages of green energy technologies such as clean, no pollution, reproducible, and environmentally friendly manner [3,4], as well as the limited supply of fossil fuels and the negative impact of carbon dioxide emissions on global climate change, necessitating the increasing

Artificial Neural Networks for Renewable Energy Systems and Real-World Applications.
DOI: https://doi.org/10.1016/B978-0-12-820793-2.00008-2

usage of renewable energy sources [5,6]. Recently, solar energy has become regarded as an effective means for addressing the impending energy crisis [7,8]. The direct method of utilizing solar energy is the solar thermal conversion method [9–12]. Among new energy systems, concentrating solar power (CSP) collectors are among the most mature technologies, and they are received considerable attention for green power generation from researchers and manufacturers [13–15] and have been used as a vital and attractive option for driving energy systems [16].

CSP has persisted in its advance into developing countries that possess high direct normal irradiance (DNI) and economical alignment and/or a clear strategy with the achievements of CSP technologies [17]. In this respect, CSP systems are currently receiving interested policy support in countries with limited gas and oil reserves, energy storage requirements, or powerful manufacturing, flexible power networks, and job creation [18]. CSP plants have the ability to generate high-temperature heat which can be used for electricity producing with power cycles and/or for meeting the heat requirements for several applications [6,19,20]. They also can be coupled with a thermal energy storage system (TES) or hybridized with renewable and nonrenewable energy sources [21–23].

Four main CSP technologies have been widely used in engineering applications. All of these are utilized to harvest solar energy by concentrating it onto a receiver. Parabolic trough collectors (PTCs) and linear Fresnel reflectors (LFRs) are line-focus types that concentrate the received sunlight onto a linear receiver, which typically is a tube that contains a working fluid [24]. The main drawback of these technologies is their limited theoretical concentration ratio. The parabolic dish and central receiver systems are point-focus types that concentrate the received sunlight onto a spot receiver with high concentration ratios [25,26]. The point-focus types require costly tracking systems to accurately follow the sun throughout the day.

In this context, the modeling of energy systems is a critical issue to establish and assess energy policies that control the energy production and consumption processes. The planning of energy policies can be established by analyzing historical energy production/consumption data to estimate their future ones [27–29]. These models can help in selecting the optimal technology among all existing ones for a certain application to meet the energy demand [30]. Moreover, it can be applied to minimize costs and maximize productivity. The industrialization of such technologies requires input–output modeling based on projected or existing production functions [31]. Therefore, defining the correct inputs and outputs is a crucial issue to develop accurate predictive models.

Artificial neural networks (ANNs) are artificial intelligence systems that mimic the organizational principles of the human brain. ANNs can learn from examples through the training process, and can also be used to figure out the relationship between the inputs and outputs of a certain system [32,33]. Therefore, ANN models have been established as robust predictive tools to handle practical engineering problems. Any engineering problem has process responses which can be determined as a function of some process input factors [34,35]. ANN models can be used to determine the correlation between the process factors and the process responses in a simpler and faster manner without involving solving complex differential equations. Therefore, the main advantages of ANN over conventional modeling techniques are simplicity, speed, and generalization capabilities in terms of learning from the examples [36].

The main features of the ANN technique, compared to other computational methods, are summarized as follows: low computational time, simplicity, high speed, and ability to solve nonlinear and complex relationships among the extracted data and the running variables [37]. On the other hand, the key limitation of the ANN technique is the deficiency of the data for training of the model, which is not the case with other analytical methods [38].

Many attempts have investigated the appliability of the ANN in diversified renewable energy applications such as photovoltaic power systems [39], solar water collectors [40], solar stills [41], wind turbines [42], greenhouses [43], solar dryers [44], storage containers [23], heat pumps [45], adsorption cooling, [46], and geothermal systems [47].

However, to the best of the authors' knowledge, there have been no previous reviews on the application of ANNs in CSP systems. For this reason, this chapter focuses on the application of ANNs in design optimization and performance modeling of different CSP systems. Different statistical standards employed to evaluate the performance of ANNs in modeling CSP collectors have been introduced. Therefore, this chapter can be regarded as of long-term scientific value for researchers that are interested in CSP technologies. This critical chapter highlights the following aspects:

- A brief discussion of the structure description and working principles of different types of CSP systems, such as PTC, solar dish collector (SDC), LFR, and central tower receiver (CTR), used in different CSP applications is introduced.

- The conceptual mechanism of ANNs, as well as several statistical performance assessment criteria employed to assess the prediction accuracy of an ANN model, have been also briefly addressed.
- The applications of ANNs in different solar concentrated collectors are summarized in detail.
- Conclusions, challenges, and recommendations for future trends are also discussed.

3.2 Concentrating solar collectors

A typical concentrating solar collector consists mainly of two parts, the concentrator and receiver, which are commercially available in many designs. Concentrators are available as reflectors or refractors, continuous or noncontinuous, and parabolic or cylindrical. Meanwhile, the receiver can be flat, concave, conical, or cylindrical, and can be with or without glazing. According to the movement of the sun during the day, positioning of the optical system is one of the crucial matters for concentrating solar collectors to achieve their best performance. Generally, concentrating solar collectors can be classified into four systems: PTCs, SDCs, LFRs, and CTRs or Heliostat field reflectors (HFRs).

3.2.1 Parabolic trough collector

The reflective material of a PTC is twisted into the parabola shape, therefore this collector is named a parabolic trough [48]. It is composed of a parabolic reflecting exterior with a receiver tube positioned along its focal line. The receiver tube made from a glass vacuum tube covered by a black metal tube is located along the focal line of the parabolic reflector, which receives the concentrated solar radiation by the reflector. Generally, the receiver is linear and its exterior is coated by a selective coating in order to increase its absorptivity and decrease the dissipated radiation. Usually, due to the positioning of the glass-coated tube through the receiver tube, the heat losses by convection are reduced and thus the temperature is increased and good efficiency could be obtained. The PTC has proved its ability to work efficiently with high heat gain and a temperature range between 50°C and 400°C [49]. Moreover, an antireflection coating is also added to improve the transmittance and to overcome transmittance losses caused by the glass tube [50]. Normally, single-axis tracking is used for a PTC, and the collector is oriented in either east–west (E–W) direction and north–south (N–S) tracking or east–west tracking and north–south direction. The N–S orientation with E–W sun tracking is more desirable because a small adjustment in the PTC is required and full aperture always pinpoints the direction of the sun at noon [51].

3.2.2 Solar dish collector

The SDC is considered as a focal point type collector as the receiver or conversion unit is installed at the focal point of the dish [16]. The solar radiation in this type is concentrated onto the receiver by a two-axis tracking system. The tracking mechanism is similar to PTC, and it looks like a satellite dish antenna arrangement as the sunlight beam is concentrated on the area where the receiver is located. The receiver absorbs solar radiation and converts it into heat energy which is transmitted to a circulating working fluid via a heat exchanger. The concentration ratio of the SDC is in the range from 600 to 2500 and the obtained temperature can reach more than 700°C due to the 2D accurate tracking system, which enables the collector to face the sun at any position [52].

3.2.3 Linear Fresnel reflector

The LFR design comprises an arrangement of linear reflecting arrays that concentrate the beam radiation to a connected linear receiver tube. In addition, a plane or elastically rounded reflector is used as a secondary reflector to enclose the top side of the absorber tube for a higher solar radiation concentration [53]. However, the concentration ratio of LFR is less than that of PTC. LFR is more economical than PTC as it has one receiver mounted to the ground and all reflectors are also fixed, which reduces the installation and mounting cost [54]. LFR has lower thermal efficiency and lower working temperatures compared to the other concentrating solar collectors, due to its low concentration ratios [53]. However, LFR is the cheapest and is applicable for an operating temperature range up to 300°C [55]. The LFR system requires a large space in order to avoid the shading effects on the reflectors, which is considered to be one of the disadvantages of this technology. However, this problem can be decreased by increasing the absorber tower height, but this solution will raise the system cost.

3.2.4 Central tower receiver

The incident solar radiation on this type of collector is reflected and concentrated on a common point (CTR) by using multiple arrays of flat mirrors to obtain a high temperature (heliostats field reflector) as shown in Fig. 3.4 [56]. In this gigantic design, a huge amount of heat can be generated which can be used for water boiling or steam generation at high pressure and temperature. The obtained heat energy can also be transmitted to an Heat Transfer Fluid (HTF) to be used for power generation or other different industrial applications [57]. The typical working fluids for CTR systems are water and molten salt. The resulting temperature from this system is up to 2000°C with concentration ratios of 300−1500, therefore, it is efficient for applications requiring a very high temperature [58].

3.3 Artificial neural networks

ANN techniques are used for modeling, simulation, optimization, and prediction of the performance of engineering systems. They have been used in different science disciplines in recent decades due to their high speed and accuracy. ANNs consist of many processing units called neurons connected by synapses [37,59]; this structure has a natural capability to figure out experimental knowledge and store it for future use.

3.3.1 Conceptual structure of artificial neural networks

ANNs have superior characteristics over other conventional modeling approaches such as high-speed information processing, fault tolerance, mapping capabilities, generalization, adaptiveness, and robustness. These characteristics support ANNs to be used as a smart and powerful tool for modeling, optimization, and prediction of the performance of various engineering systems. Therefore, they have been utilized to solve complex nonlinear problems in real-world applications with a significant reduction in time and cost without requiring the solving of complex mathematical models [60,61]. In general, the neuron-based architecture used in establishing various ANN models consists of many connecting links called synapses; each synapse has its synaptic weight, w_{kj} (as in Fig. 3.1). Each synaptic weight is multiplied by its own input xj and then all weighted inputs are summed. To control the summation's output v_k an external bias

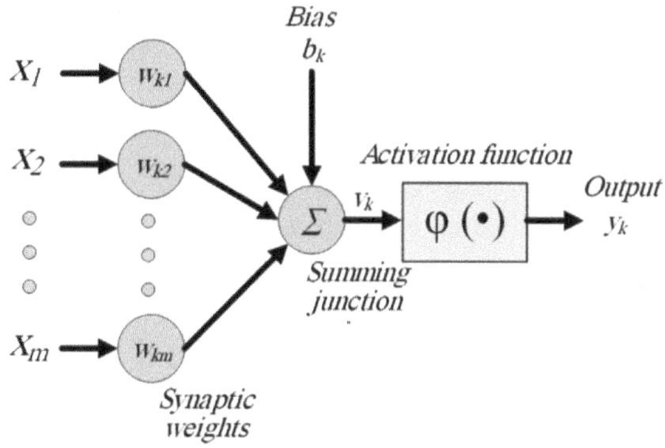

Figure 3.1 A nonlinear model of a neuron.

b_k is added to lower or increase the output value. Then the final output is subjected to an activation function φ to reduce the amplitude of the output signal y_k into a finite value; these mathematical operations can be expressed as follows:

$$y_k = \varphi \left[\sum_{j=1}^{m} (\omega_{jk}) + b_k \right] \tag{3.1}$$

where j and k denote the synapse and neuron number, respectively.

3.3.2 Performance evaluation criteria of the artificial neural network model

The evaluation of ANN model performance is dependent on the selection criteria of minimal errors of several statistical measures such as mean absolute error (MAE), root mean square error (RMSE), coefficient of variance (COV), and best fit of ANN predicted data with actual available data based on regression coefficient (R^2), which they can be computed by Eqs. (3.2)–(3.5) as follows [62–64]:

$$MAE = \frac{1}{n} \sum_{i=1}^{n} (X_a - X_b) \tag{3.2}$$

$$RMSE = \sqrt{\frac{1}{n} \sum_{i=1}^{n} (X_a - X_b)^2} \tag{3.3}$$

$$COV = \frac{RMSE}{\frac{1}{n} \sum_{i=1}^{n} X_b} \times 100 \tag{3.4}$$

$$R^2 = 1 - \frac{\sum_{i=1}^{n} (X_a - X_b)^2}{\sum_{i=1}^{n} X_b^2} \quad (3.5)$$

where, n, x_a, and x_b are the number of experiments, the experimental value, and the predicted value, respectively.

3.4 Artificial neural network applications in concentrating solar power systems

The following text presents a comprehensive state-of-the-art review of the published studies concerned with the applications of ANNs in various CSP systems, namely PTC, LFR, CTR, and SDC.

Many attempts have been made to improve the design structure and maximize the thermal performance of the PTC system as well as minimize the high cost required for construction and testing of the PTC field by using numerical analysis to predict the outputs from input variables using the different approaches of ANN models. The ANN technique is considered to be an effective solution for this purpose as it showed acceptable prediction, especially for the nonlinear nature of the training data [65].

Initially, one of the first efforts conducted on the use of ANNs for PTC modeling was carried out by Kalogirou [66], who used an ANN to model a solar steam generator powered by PTC at different conditions and various aspects. The system consisted of a PTC, flash vessel, and high-pressure circulating pump, with all parts connected through a pipe network as shown in Fig. 3.2. Using ANN facilitated handling this problem which was very difficult to do with analytical methods as the system operates under transient conditions. The results showed that the HTF temperature profile at different points of the system could be predicted within 3.9% as an RMSE, which was regarded as tolerable for PTC design.

Vaferi et al. [67] developed a forward feed ANN (FFANN) model based on a multilayered perceptron (MLP) for evaluating the heat transfer performance enhancement of CuO/water, Al_2O_3/water, and SiO_2/water nanofluids in cylindrical pipes for PTC systems. The inputs and outputs of the model are clearly shown in Fig. 3.3. This model indicated that the R^2 value was 0.9997, and the RMSE was 1.7×10^{-5} for the predicted coefficient of convection heat transfer.

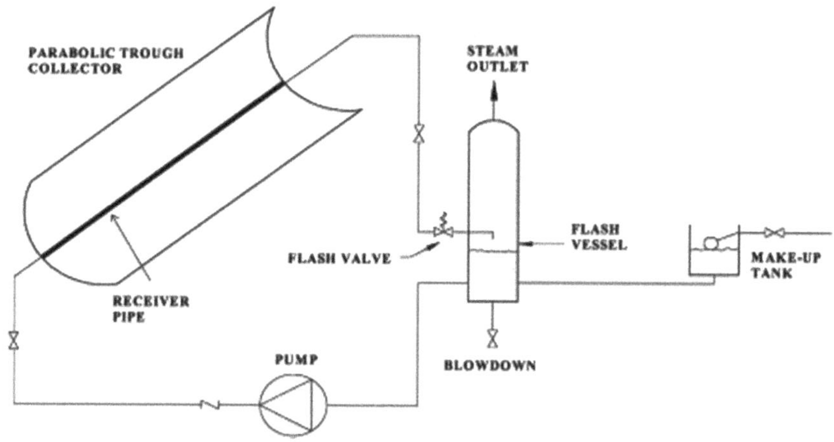

Figure 3.2 The steam production device powered by PTC.

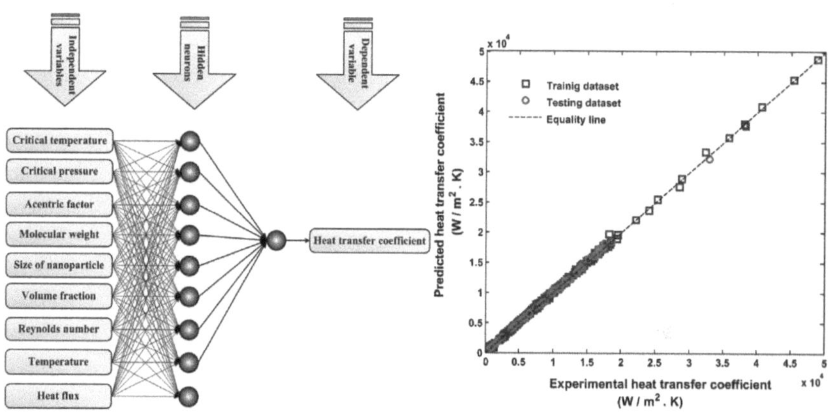

Figure 3.3 (A) The inputs and outputs of the model; (B) compression of the ANN predicting results and experimental data for the heat transfer coefficient [67].

Heng et al. [68] predicted the exit fluid temperature of a PTC tube using the ANN technique. The model was trained using simulation results obtained by the superposition principle combined with the finite element method (FEM). The inputs of the model were DNI (0–6000 W/m^2), HTF velocity (0.3–1.5 m/s), and time interval (1200–5 seconds). The daily outlet HTF temperature was predicted within 1 minutes of computational time with Root Mean Average Error (RMAE) less than 2K. The outlet HTF temperature of the tube of PTC for one year of working can be estimated in less than 6 hours. The predicted findings can be regarded as primary system design, thermodynamic balance analysis, and system planning.

In order to minimize the entropy generation of Al_2O_3-ethylene glycol (EG)/water nanofluid flow inside a PTC tube under turbulent flow regime, ANN modeling and quadratic algorithm were used by Moghadam et al. [69], as clearly illustrated in Fig. 3.4. The nanoparticle diameter, nanofluid average flow temperature, and Reynolds number (Re) have been employed as three ANN inputs to investigate the optimal volume fraction. It was found that when the volume fraction was decreased, and the nanoparticle diameter and average flow temperature of the Al_2O_3-EG/water nanofluid were increased, the entropy generation was remarkably decreased. It was also reported that the optimal values of Re, nanoparticle diameter, and the average temperature of nanofluids were 4000, 90 nm, and 87°C, respectively.

An ANN model has been applied by Tzuc et al. [70] for the thermal performance optimization of PTC. For that purpose, an FFANN approach was applied with experimental results from PTC testings. In order to estimate and compare the simulated results of the thermal performance with the value of $R^2 = 0.9996$ with experimental results, many different parameters such as the HTF temperature inlet and outlet, water flow rate, wind velocity, rim angle, direct solar radiation, and ambient temperature were used as the main input variables. The results showed that ANNs could be used as a powerful tool for PTC performance optimization.

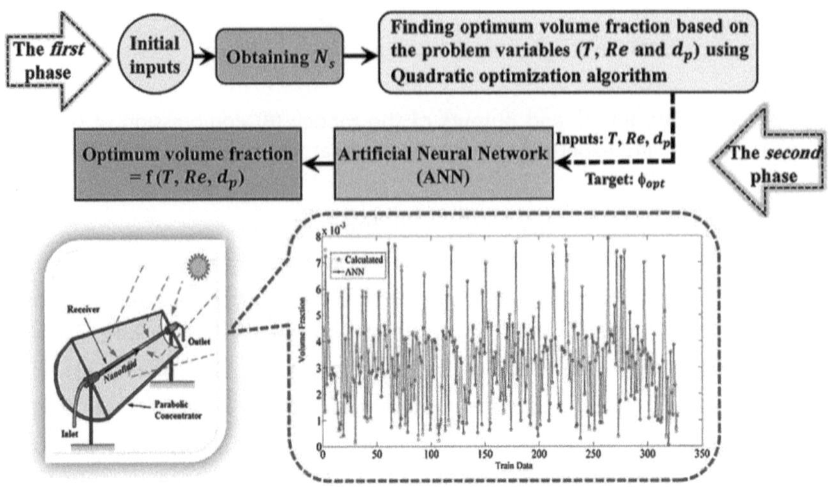

Figure 3.4 Representation of the optimum nanofluid conditions using ANN [69].

Based on outdoor experimental data of the parabolic trough field, ANN was applied by Kalogirou et al. [71] to predict the PTC energy efficiency. The results showed a good response of the system, with an R^2 value of 0.9999 and the maximum deviation between the actual and predicted values was less than 3.9%. It was also concluded that under the different input parameters a good match between the real and predicted performances was achieved.

The thermal efficiency of a solar PTC was experimentally carried out and theoretically investigated by Tafarroj et al. [72] using the ANN-MLP model. SiO_2 and multiwalled carbon nanotube (MWCNT) nanofluids with two different volume fractions were used as an HTF. The results indicated that the maximum obtained outlet temperature was 74°C when a 0.6 wt.% MWCNT/EG nanofluid was used as an HTF. The results also showed that the trained MLP network could provide a more accurate prediction for the process outputs with less time and cost.

Tellez et al. [73] developed an ANN-MLP model and its inverse with low-cost components for PTC to predict the outlet temperature of hot water and optimize the system performance. The experimental and predicted data were compared and the comparison results showed an acceptable agreement (RMSE $>$ 0.805, $R^2 > 0.985$, and MAE $\sim$0.059). Moreover, the inverse ANN and genetic algorithms (GAs) solution methods also were used for optimization of the feeding tank temperature to increase the water outlet temperature. It was reported that the highest predicted outlet temperature was 49°C.

In another study, an ANN-MLP model was developed by Boukelia et al. [74] to determine the best way for the performance prediction and techno-economic optimization of solar parabolic trough power plant (SPTPP) hybridized with fuel recovery system and TES. The estimated statistical measures unveiled that the ANN-MLP algorithm with 38 neurons in a single hidden layer showed the best ANN model to predict the yearly accumulated power production and levelized electricity cost of the proposed SPTPP. The results showed that a minimum levelized electricity cost of 0.088 \$/kWh for the presented SPTPP could be achieved at the design ambient temperature and solar radiation of 34°C and 850 W/m^2, respectively, for 2.5 hours for the storage system while the solar multiple and the row spacing were 1.7 and 23 m, respectively.

Solar dish systems recently have been considered as one of the most promising technologies for co-generating heat and power production. By using these systems a huge amount of solar energy can be concentrated, consequently its temperature increases to very high levels and thereby larger heat capacities could be established for freshwater and power production. A few attempts have investigated the potential application of ANN models to predict the design criteria and thermal performance parameters of solar dish technology.

Loni et al. [75] applied the Data Handling Group Method (DHGM) as an ANN technique to investigate and predict both the thermal efficiency and cavity heat gain of a hemispherical cavity receiver integrated with an SDC. The ambient temperature, direct solar irradiance, and outlet temperature were used as input variables. It was reported that the obtained R^2 values for the proposed model with the input variables for the thermal efficiency and heat gain of the receiver cavity based on the proposed model were 0.9567 and 0.9709, respectively.

In another study, Loni et al. [76] applied a numerical modeling method and an ANN-based method as two different methods to investigate the thermal performance of an SDC with a rectangular-tubular cavity receiver. The thermal oil was utilized as the HTF in the SDC system. The solar irradiance, the cavity depth, and the diameter of the tube were considered as variable parameters for the ANN model, whereas the thermal performance was the output. The comparative results for the numerical method and ANN model showed that a remarkable agreement with $R^2 = 0.999$ has been achieved at the different variable parameters. Moreover, it was concluded that the ANN-based method is considered to be an effective tool to save the commutation time and cost required to determine the optimal structure of the cavity receiver.

Recently, Khosravi et al. [77] developed a novel optimization model based on five different hybrid intelligent network techniques, Adaptive Neuro-Fuzzy Inference System (ANFIS), DHGM, ANN-MLP, GA, and Particle Swarm Optimization (PSO), to determine the impact of different operative and geometrical variables on the generated power and efficiency of the Solar Dish Stirling System (SDSS), taking Natal city, Brazil as a case study, as presented in Fig. 3.5. The findings unveiled that the intelligent approaches can be considered as an efficient alternative for energy modeling of the SDSS and the statistical matrices showed that the ANFIS-PSO method performed better than the other studied approaches, as it optimized the overall efficiency of the SDSS by up to 35%.

An FFANN model optimized by the GA was developed by Sameti et al. [78] to determine the required power for an SDSS in a smart grid. It was found that accurate results for the performance could be obtained

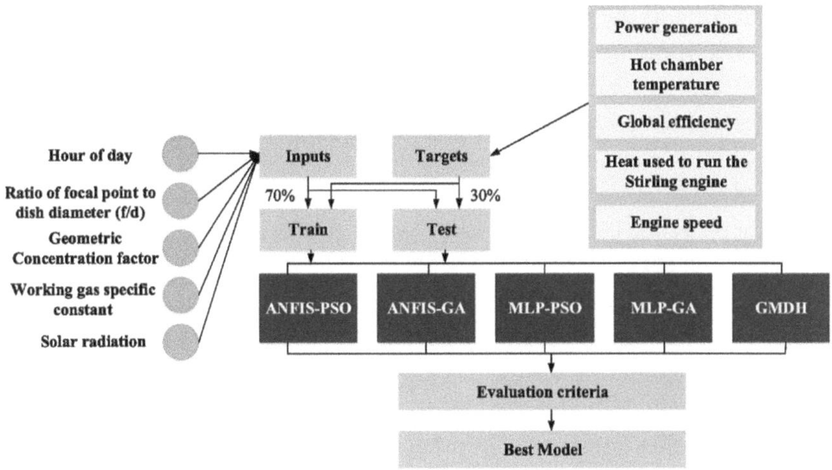

Figure 3.5 The novel intelligent networks for energy modeling of the SDSS [77].

for R^2 values greater than 0.9, while acceptable results could be achieved at the range of 0.8–0.9 for the R^2 value. It has been reported that future research is concerned with the optimization of the ANN structure through an appropriate GA method.

In order to predict the effect of the various parameters of the performance of SDC on the heat losses of the receiver, an ANN model was developed by Vikram and Reddy [79]. They also investigated the effects of different parameters such as operating temperature, insulation thickness, the angle of inclination, diameter ratio, and emissivity of the cavity cover on the heat losses from the modified cavity receiver. The results showed that the total heat loss of the receiver was 25% for receiver inclination angles of 0 and 90 degrees. Besides, the diameter ratio also has a significant effect on the heat losses of the receiver cavity. Furthermore, it was concluded that the proposed ANN method proved its ability to predict heat losses better than any previously existing model.

Some researchers have applied various models of ANN to predict the design considerations, performance outputs, and economical matrices of the CTR CSP system, as it is considered the most mature and most used in CSP plants.

Lopez et al. [80] applied an ANN model for a solar power tower to study the effect of water vapor on the solar radiation extinction between heliostats and the receiver. For this purpose, an atmospheric transmission code (MODTRAN) was utilized to simulate the values of DNI arriving at the mirror and the receiver with various geometries and under different atmospheric conditions associated with water vapor and aerosols. After that, the resulting variables were used as inputs to an ANN to estimate the corresponding DNI extinction. The proposed ANN modeling was able to provide the DNI attenuation over a wide range of input variables, with RMSE differences of only 0.8%.

Khosravi et al. [81] introduced thermodynamic-economical modeling of a 100 MW CTR power plant with molten salt storage for five different cities, New Delhi, Thanjavur, Islamabad, Kabul, and Dhaka, as seen in Fig. 3.6. An innovative hybrid approach based on optimized ANFIS with the implementation of GA and a teaching–learning-based optimization algorithm (GATLBO) was developed. The input parameters were longitude, latitude, DNI, and solar multiple, while the yearly generated power, energy cost, and factor of capacity were the output targets. The results revealed that the yearly energy generated by the CTR station increased by increasing the solar multiple and reducing the DNI design point. It was addressed that the ANFIS-GATLBO technique efficiently predicted the outputs with R^2 close to 1.

Yaïci et al. [82] proposed ANN-MLP modeling for evaluating the performance behavior parameters of a CTR energy system. Several comparisons with real explanatory data from a CTR system tested in Ottawa, Canada, over 2 years under various atmospheric conditions were performed. The predicted results demonstrated the high prediction accuracy attainable with each of the models using reduced input variables.

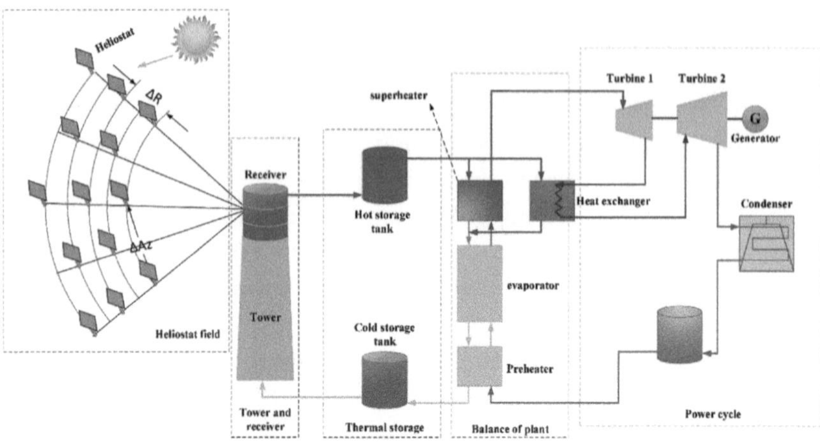

Figure 3.6 A schematic of CTR power plant [81].

A new method to optimize the performance of CTR salt cavity receivers considering thermal, optical, operational, and hydraulic aspects was proposed by Schött et al. [83]. The hydraulic layout, the receiver geometry, and the aiming strategy were simultaneously optimized. ANN was applied to predict the total pressure loss and the annual yield of the system, as shown in Fig. 3.7. The objective function was established based on a validated simulation model incorporated with a sky discretization approach. The proposed methodology was applied to employ a 55 MW_{th} receiver and maximize its performance. The annual yield was increased by about 4% using the optimized receiver configuration compared with that of the reference configuration.

With regard to the LFR CSP systems, there are limited studies that have been reported in the literature where the objective was the optimization of an LFR in order to predict their performance with low cost and time. Zhu [84,85] developed an adaptive optimization approach for the secondary reflector of the LFR system shown in Fig. 3.8. The adaptive approach began

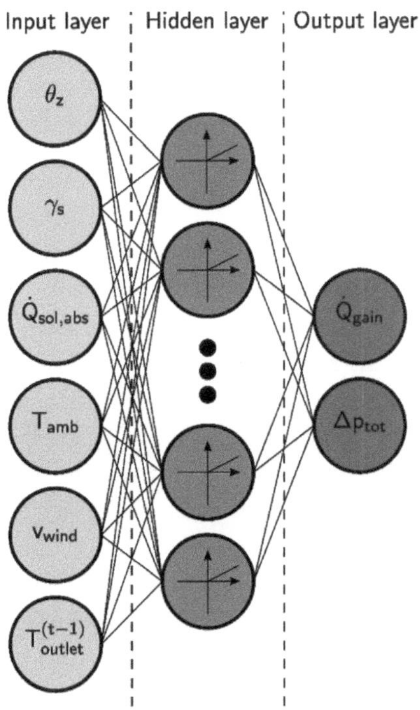

Figure 3.7 ANN input layers, one hidden layer with 100 *ReLU* neurons, and output layers [83].

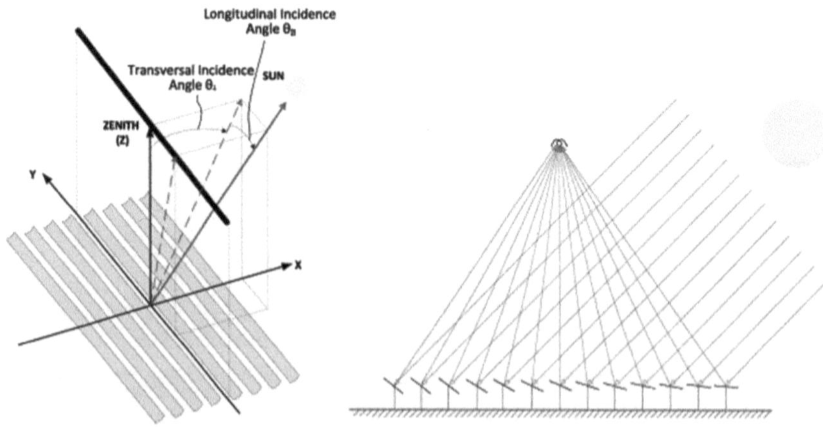

Figure 3.8 Schematics of a typical LFR [85].

at a single edge point and built the next surface point in an adaptive form. The purpose of the approach was to maximize the reflected radiation from Fresnel reflectors to the receiver tube. It was observed that the incidence angle modifier indication with regard to independent longitudinal and transversal components can be a satisfactory approach, however, it relatively reduced the optical performance of a LFR.

3.5 Prospective and challenges

ANN is an artificial intelligence-based numerical tool that has been applied to solve nonlinear multivariable problems related to complex physical systems; such systems are difficult to solve using conventional modeling techniques and require many assumptions. More specifically, in the CSP devices applications, ANN-based methods have been successfully utilized to evaluate the effects of various operating factors as well as solar radiation variables on the device performance. They also have been applied as a robust predictive tool with better accuracy compared with conventional regression models.

ANN has many crucial advantages that make them the most appropriate techniques to model various engineering systems:

1. ANN has the ability to model, figure out, and learn complex nonlinear relationships between the process factors and responses, which is very important because, in real-world applications, most engineering systems have complex nonlinear nature.

2. ANN has excellent generalization capabilities. After the learning process from the initial inputs/outputs is accomplished, it can infer the relationships in the data that it has not seen before.
3. Unlike some other prediction methods, ANN does not have any restrictions on the process variables. Furthermore, it can learn hidden relationships between inputs/outputs of a certain system without imposing any fixed relationships in the data. This may be very useful in the forecasting of meteorological time series with high data volatility.

3.6 Conclusions and future recommendations

CSP plants are considered a promising solution toward meeting the worldwide growing demand for energy production. In this chapter, the current status and recent progress in the performance prediction of CSP systems using ANNs have been demonstrated. The summarized data for several kinds of CSP systems such as parabolic troughs, linear Fresnel, parabolic dish, and solar tower CSP power plants have been utilized for training of ANN models with the employment of diversified types of learning algorithms and to explore the optimal ANN modeling that can be successfully applied to predict the performance of the CSP system.

An ANN model is structured with three layers: an input layer, an output layer, and a hidden layer. The parameters that directly affect the system performance are selected as input parameters in the input layer. In the output layer, the estimated data are selected for output parameters. The hidden layers are processing elements, in which optimum numbers of neurons are taken. Regarding the reviewed papers, which include the latest research work in the applications of ANN for modeling different types of CSP systems, the following key conclusions can be made:

- The reviewed ANN models allow performance prediction and design characterization for the different types of CSP technologies to be done with a great level of accuracy. Furthermore, since most of them are based on operating and opt-geometrical parameters it is possible to estimate their performance behavior at the desired location if the operating and opt-geometrical parameters are available.
- Most of the applied ANN models have focused on performance prediction and economical assessment of the solar parabolic trough system and solar CTR CSP systems. Conversely, very limited studies have

investigated the potential application models to evaluate the feasibility and thermodynamic performance of solar dish and LFR CSP systems.

- Different types of ANN models, in particular Multilayered Perceptron (MLP), Forward Feed ANN (FFANN), DHGM, and ANFIS models are used for performance prediction of CSP systems.
- The ANN-MLP model is the most widely used and well-developed model for predicting the performance of CSP devices, and is an excellent alternative tool for the simulation of such a CSP system with reduced computation time.

However, although ANNs have been utilized to model and solve several problems related to the CSP technologies and they have succeeded in obtaining excellent results, some important issues related to these technologies have been not investigated yet. Some of these important issues in which the ANNs could be applied are as follows:

- The application of ANNs in weather installation, thermodynamic performance prediction, and techno-economic analyses of solar dish and LFR-powered CSP power plants are still worthy of in-depth study.
- Another important further aspect should be also recommended to be focused on the performance prediction and development of different innovative multigeneration-based CSP systems using advanced principles of thermodynamics based on ANN models such as thermo-economic, exergo-environmental, and exergo-economic in order to discover the thermal, environmental, and economic assessments of the novel developed multigeneration systems.
- The dual combination of ANN models hybridized with advanced optimization techniques may be regarded as a promising hybrid ANN model to augment the flexibility and accuracy of the applied learning algorithms. Such hybrid ANN models can help to determine the optimal parameters of ANNs to optimize the performance of different CSPs to become more cost-effective and efficient in the future.
- Comparative studies of different ANN learning algorithms based on performance predictions of CSP technologies are greatly needed to discover the optimal learning algorithms.

References

[1] L. Jing, J. Zhao, H. Wang, W. Li, Y. Du, Q. Zhu, et al., Numerical analysis of the effect of swirl angle and fuel equivalence ratio on the methanol combustion characteristics in a swirl burner, Process Safety and Environmental Protection 158 (2022) 320–330.

[2] J. Zhao, L. Ma, M.E. Zayed, A.H. Elsheikh, W. Li, Q. Yan, et al., Industrial reheating furnaces: a review of energy efficiency assessments, waste heat recovery potentials, heating process characteristics and perspectives for steel industry, Process Safety and Environmental Protection 147 (2021) 1209−1228.

[3] A.E. Kabeel, A. Khalil, S.M. Shalaby, M.E. Zayed, Experimental investigation of thermal performance of flat and v-corrugated plate solar air heaters with and without PCM as thermal energy storage, Energy Conversion and Management 113 (2016) 264−272.

[4] M.E. Zayed, J. Zhao, Y. Du, A.E. Kabeel, S.M. Shalaby, Factors affecting the thermal performance of the flat plate solar collector using nanofluids: a review, Solar Energy 182 (2019) 382−396.

[5] S.A. El-Agouz, M.E. Zayed, A.M. Abo Ghazala, A.R.A. Elbar, M. Shahin, M.Y. Zakaria, K.K. Ismaeil, Solar thermal feed preheating techniques integrated with membrane distillation for seawater desalination applications: Recent advances, retrofitting performance improvement strategies, and future perspectives, Process Safety and Environmental Protection 164 (2022) 595−612.

[6] D. Oliva, M.A. Elaziz, A.H. Elsheikh, A.A. Ewees, A review on *meta*-heuristics methods for estimating parameters of solar cells, Journal of Power Sources 435 (2019) 126683.

[7] A.E. Kabeel, A. Khalil, S.M. Shalaby, M.E. Zayed, Improvement of thermal performance of the finned plate solar air heater by using latent heat thermal storage, Applied Thermal Engineering 123 (2017) 546−553.

[8] F.A. Hammad, S.M. Shalaby, A.E. Kabeel, M.E. Zayed, Experimental investigation and thermo-economic performance analysis of a modified solar distiller design with thermal storage material and v-corrugated absorber basin, Journal of Energy Storage 52 (2022) 105020.

[9] L. Ma, Y. Li, J. Wang, S. Li, J. Zhao, W. Li, et al., A thermal-dissipation correction method for in-situ soil thermal response test: experiment and simulation under multi-operation conditions, Energy and Buildings 194 (2019) 218−231.

[10] A.H. Elsheikh, S.W. Sharshir, M.K. Ahmed Ali, J. Shaibo, E.M.A. Edreis, T. Abdelhamid, et al., Thin film technology for solar steam generation: a new dawn, Solar Energy 177 (2019) 561−575.

[11] F.A. Essa, A.H. Elsheikh, A.A. Algazzar, R. Sathyamurthy, M.K. Ahmed Ali, M.A. Elaziz, et al., Eco-friendly coffee-based colloid for performance augmentation of solar stills, Process Safety and Environmental Protection (2020).

[12] D. Mevada, H. Panchal, M. Ahmadein, M.E. Zayed, N.A. Alsaleh, J. Djuansjah, et al., Investigation and performance analysis of solar still with energy storage materials: an energy- exergy efficiency analysis, Case Studies in Thermal Engineering 29 (2022) 101687.

[13] M.E. Zayed, J. Zhao, A.H. Elsheikh, Y. Du, F.A. Hammad, L. Ma, et al., Performance augmentation of flat plate solar water collector using phase change materials and nanocomposite phase change materials: a review, Process Safety and Environmental Protection 128 (2019) 135−157.

[14] A.H. Elsheikh, S.W. Sharshir, M.E. Mostafa, F.A. Essa, M.K. Ahmed, Ali, Applications of nanofluids in solar energy: a review of recent advances, Renewable and Sustainable Energy Reviews 82 (2018) 3483−3502.

[15] M.E. Zayed, J. Zhao, A.H. Elsheikh, F.A. Hammad, L. Ma, Y. Du, et al., Applications of cascaded phase change materials in solar water collector storage tanks: a review, Solar Energy Materials and Solar Cells 199 (2019) 24−49.

[16] M.E. Zayed, J. Zhao, W. Li, A.H. Elsheikh, Z. Zhao, A. Khalil, et al., Performance prediction and techno-economic analysis of solar dish/stirling system for electricity generation, Applied Thermal Engineering 164 (2020) 114427.

[17] M.E. Zayed, J. Zhao, A.H. Elsheikh, Z. Zhao, S. Zhong, A.E. Kabeel, Comprehensive parametric analysis, design and performance assessment of a solar dish/Stirling system, Process Safety and Environmental Protection 146 (2021) 276–291.
[18] A.G. Fernández, J. Gomez-Vidal, E. Oró, A. Kruizenga, A. Solé, L.F. Cabeza, Mainstreaming commercial CSP systems: a technology review, Renewable Energy 140 (2019) 152–176.
[19] A.H. Elsheikh, S.W. Sharshir, M. Abd Elaziz, A.E. Kabeel, W. Guilan, Z. Haiou, Modeling of solar energy systems using artificial neural network: a comprehensive review, Solar Energy 180 (2019) 622–639.
[20] M.M. Aboelmaaref, M.E. Zayed, J. Zhao, W. Li, A.A. Askalany, M. Salem Ahmed, et al., Hybrid solar desalination systems driven by parabolic trough and parabolic dish CSP technologies: technology categorization, thermodynamic performance and economical assessment, Energy Conversion and Management 220 (2020) 113103.
[21] A.E. Kabeel, A. Khalil, S.M. Shalaby, M.E. Zayed, Investigation of the Thermal Performances of Flat, Finned, and v-Corrugated Plate Solar Air Heaters, Journal of Solar Energy Engineering 138 (2016).
[22] S.M. Shalaby, A. Khalil, A.E. Kabeel, M.E. Zayed, Improvement of the thermal performance of the V-corrugated plate solar air heater with pcm by using insulated upper cover during night, 2018 IEEE International Conference on Smart Energy Grid Engineering (SEGE), IEEE, 2018, pp. 346–350.
[23] M.E. Zayed, J. Zhao, W. Li, A.H. Elsheikh, A.M. Elbanna, L. Jing, et al., Recent progress in phase change materials storage containers: geometries, design considerations and heat transfer improvement methods, Journal of Energy Storage 30 (2020) 101341.
[24] Y.-L. He, Y. Qiu, K. Wang, F. Yuan, W.-Q. Wang, M.-J. Li, et al., Perspective of concentrating solar power, Energy 198 (2020) 117373.
[25] M.T. Islam, N. Huda, A.B. Abdullah, R. Saidur, A comprehensive review of state-of-the-art concentrating solar power (CSP) technologies: current status and research trends, Renewable and Sustainable Energy Reviews 91 (2018) 987–1018.
[26] Y. Wang, W. Lipiński, J. Pye, A method for in situ measurement of directional and spatial radiosity distributions from complex-shaped solar thermal receivers, Solar Energy 201 (2020) 732–745.
[27] A. Silvast, E. Laes, S. Abram, G. Bombaerts, What do energy modellers know? An ethnography of epistemic values and knowledge models, Energy Research & Social Science 66 (2020) 101495.
[28] R. Alizadeh, L. Soltanisehat, P.D. Lund, H. Zamanisabzi, Improving renewable energy policy planning and decision-making through a hybrid MCDM method, Energy Policy 137 (2020) 111174.
[29] M.E. Zayed, J. Zhao, W. Li, A.H. Elsheikh, M.A. Elaziz, D. Yousri, et al., Predicting the performance of solar dish Stirling power plant using a hybrid random vector functional link/chimp optimization model, Solar Energy 222 (2021) 1–17.
[30] M.E. Zayed, J. Zhao, A.H. Elsheikh, W. Li, M.A. Elaziz, Optimal design parameters and performance optimization of thermodynamically balanced dish/Stirling concentrated solar power system using multi-objective particle swarm optimization, Applied Thermal Engineering 178 (2020) 115539.
[31] H.K. Ozturk, H. Ceylan, A. Hepbasli, Z. Utlu, Estimating petroleum exergy production and consumption using vehicle ownership and GDP based on genetic algorithm approach, Renewable and Sustainable Energy Reviews 8 (2004) 289–302.
[32] H.K. Ghritlahre, R.K. Prasad, Application of ANN technique to predict the performance of solar collector systems—a review, Renewable and Sustainable Energy Reviews 84 (2018) 75–88.

[33] M. Abd Elaziz, S. Senthilraja, M.E. Zayed, A.H. Elsheikh, R.R. Mostafa, S. Lu, A new random vector functional link integrated with mayfly optimization algorithm for performance prediction of solar photovoltaic thermal collector combined with electrolytic hydrogen production system, Applied Thermal Engineering 193 (2021) 117055.
[34] O.I. Abiodun, A. Jantan, A.E. Omolara, K.V. Dada, N.A. Mohamed, H. Arshad, State-of-the-art in artificial neural network applications: a survey, Heliyon 4 (2018) e00938.
[35] M.E. Zayed, J. Zhao, W. Li, A.H. Elsheikh, M.A. Elaziz, A hybrid adaptive neuro-fuzzy inference system integrated with equilibrium optimizer algorithm for predicting the energetic performance of solar dish collector, Energy 235 (2021) 121289.
[36] O.E. Agwu, J.U. Akpabio, S.B. Alabi, A. Dosunmu, Artificial intelligence techniques and their applications in drilling fluid engineering: a review, Journal of Petroleum Science and Engineering 167 (2018) 300–315.
[37] S. Haykin, Neural Networks: A Comprehensive Foundation, Prentice Hall PTR, 1994.
[38] R. Lippmann, Book Review: "Neural Networks, A Comprehensive Foundation," by Simon Haykin, International Journal of Neural Systems 5 (1994) 363–364.
[39] S. Ferlito, G. Adinolfi, G. Graditi, Comparative analysis of data-driven methods online and offline trained to the forecasting of grid-connected photovoltaic plant production, Applied Energy 205 (2017) 116–129.
[40] M. Souliotis, S. Kalogirou, Y. Tripanagnostopoulos, Modelling of an ICS solar water heater using artificial neural networks and TRNSYS, Renewable Energy 34 (2009) 1333–1339.
[41] F.A. Essa, M. Abd Elaziz, A.H. Elsheikh, An enhanced productivity prediction model of active solar still using artificial neural network and Harris Hawks optimizer, Applied Thermal Engineering 170 (2020) 115020.
[42] K. Karabacak, N. Cetin, Artificial neural networks for controlling wind–PV power systems: a review, Renewable and Sustainable Energy Reviews 29 (2014) 804–827.
[43] M. Taki, Y. Ajabshirchi, S.F. Ranjbar, A. Rohani, M. Matloobi, Heat transfer and MLP neural network models to predict inside environment variables and energy lost in a semi-solar greenhouse, Energy and Buildings 110 (2016) 314–329.
[44] B.K. Bala, M.A. Ashraf, M.A. Uddin, S. Janjai, Experimental and neural network prediction of the performance of a solar tunnel drier for drying jackfruit bulbs and leather, Journal of Food Process Engineering 28 (2005) 552–566.
[45] H. Esen, M. Esen, O. Ozsolak, Modelling and experimental performance analysis of solar-assisted ground source heat pump system, Journal of Experimental & Theoretical Artificial Intelligence 29 (2017) 1–17.
[46] M.E. Zayed, V.P. Katekar, R. Kumar Tripathy, S.S. Deshmukh, A.H. Elsheikh, Predicting the yield of stepped corrugated solar distiller using kernel-based machine learning models, Applied Thermal Engineering (2022) 118759.
[47] G. Liu, B. Zhou, S. Liao, Inverting methods for thermal reservoir evaluation of enhanced geothermal system, Renewable and Sustainable Energy Reviews 82 (2018) 471–476.
[48] L. Kumar, M. Hasanuzzaman, N.A. Rahim, Global advancement of solar thermal energy technologies for industrial process heat and its future prospects: a review, Energy Conversion and Management 195 (2019) 885–908.
[49] Y. Krishna, M. Faizal, R. Saidur, K.C. Ng, N. Aslfattahi, State-of-the-art heat transfer fluids for parabolic trough collector, International Journal of Heat and Mass Transfer 152 (2020) 119541.
[50] A. Sandá, S.L. Moya, L. Valenzuela, Modelling and simulation tools for direct steam generation in parabolic-trough solar collectors: a review, Renewable and Sustainable Energy Reviews 113 (2019) 109226.

[51] İ.H. Yılmaz, A. Mwesigye, Modeling, simulation and performance analysis of parabolic trough solar collectors: a comprehensive review, Applied Energy 225 (2018) 135–174.
[52] M.E. Zayed, J. Zhao, A.H. Elsheikh, W. Li, S. Sadek, M.M. Aboelmaaref, A comprehensive review on Dish/Stirling concentrated solar power systems: design, optical and geometrical analyses, thermal performance assessment, and applications, Journal of Cleaner Production 283 (2021) 124664.
[53] S. Suman, M.K. Khan, M. Pathak, Performance enhancement of solar collectors—a review, Renewable and Sustainable Energy Reviews 49 (2015) 192–210.
[54] E. Bellos, Progress in the design and the applications of linear Fresnel reflectors—a critical review, Thermal Science and Engineering Progress 10 (2019) 112–137.
[55] A. Kasaeian, S. Tabasi, J. Ghaderian, H. Yousefi, A review on parabolic trough/Fresnel based photovoltaic thermal systems, Renewable and Sustainable Energy Reviews 91 (2018) 193–204.
[56] M. Imtiaz Hussain, C. Ménézo, J.-T. Kim, Advances in solar thermal harvesting technology based on surface solar absorption collectors: a review, Solar Energy Materials and Solar Cells 187 (2018) 123–139.
[57] M.E. Zayed, A.H. Elsheikh, F.A. Essa, A.M. Elbanna, W. Li, J. Zhao, Chapter 12—High-temperature solar selective absorbing coatings for concentrated solar power systems, in: K.Y. Cheong, A. Apblett (Eds.), Sustainable Materials and Green Processing for Energy Conversion, Elsevier, 2022, pp. 361–398.
[58] A.L. Avila-Marin, J. Fernandez-Reche, A. Martinez-Tarifa, Modelling strategies for porous structures as solar receivers in central receiver systems: a review, Renewable and Sustainable Energy Reviews 111 (2019) 15–33.
[59] H.A. Babikir, M.A. Elaziz, A.H. Elsheikh, E.A. Showaib, M. Elhadary, D. Wu, et al., Noise prediction of axial piston pump based on different valve materials using a modified artificial neural network model, Alexandria Engineering Journal 58 (2019) 1077–1087.
[60] D.B. Jani, M. Mishra, P.K. Sahoo, Application of artificial neural network for predicting performance of solid desiccant cooling systems—a review, Renewable and Sustainable Energy Reviews 80 (2017) 352–366.
[61] E. Siami-Irdemoosa, S.R. Dindarloo, Prediction of fuel consumption of mining dump trucks: a neural networks approach, Applied Energy 151 (2015) 77–84.
[62] W. Yaïci, E. Entchev, Adaptive Neuro-Fuzzy Inference System modelling for performance prediction of solar thermal energy system, Renewable Energy 86 (2016) 302–315.
[63] M. Zayed, J. Zhao, W. Li, A. Elsheikh, A. Elbanna, J. Liu, et al., Recent progress in phase change materials storage containers: geometries, design considerations and heat transfer improvement methods, The Journal of Energy Storage (2020).
[64] T.A. Shehabeldeen, M.A. Elaziz, A.H. Elsheikh, J. Zhou, Modeling of friction stir welding process using adaptive neuro-fuzzy inference system integrated with harris hawks optimizer, Journal of Materials Research and Technology 8 (2019) 5882–5892.
[65] F. Almonacid, E.F. Fernandez, A. Mellit, S. Kalogirou, Review of techniques based on artificial neural networks for the electrical characterization of concentrator photovoltaic technology, Renewable and Sustainable Energy Reviews 75 (2017) 938–953.
[66] S.A. Kalogirou, Artificial neural networks in renewable energy systems applications: a review, Renewable and Sustainable Energy Reviews 5 (2001) 373–401.
[67] B. Vaferi, F. Samimi, E. Pakgohar, D. Mowla, Artificial neural network approach for prediction of thermal behavior of nanofluids flowing through circular tubes, Powder Technology 267 (2014) 1–10.
[68] S.Y. Heng, Y. Asako, T. Suwa, K. Nagasaka, Transient thermal prediction methodology for parabolic trough solar collector tube using artificial neural network, Renewable Energy 131 (2019) 168–179.
[69] A. Ebrahimi-Moghadam, B. Mohseni-Gharyehsafa, M. Farzaneh-Gord, Using artificial neural network and quadratic algorithm for minimizing entropy generation of

Al_2O_3-EG/W nanofluid flow inside parabolic trough solar collector, Renewable Energy 129 (2018) 473–485.

[70] O.M. Tzuc, A. Bassam, M.A.E. Soberanis, E. Venegas-Reyes, O.A. Jaramillo, L.J. Ricalde, et al., Modeling and optimization of a solar parabolic trough concentrator system using inverse artificial neural network, Journal of Renewable and Sustainable Energy 9 (2017) 013701.

[71] S. Kalogirou, C. Neocleous, C. Schizas, Artificial neural networks in modelling the Heat-up response of a solar steam generating plant, Proceedings of International Conference EANN'96 (1996).

[72] M.M. Tafarroj, R. Daneshazarian, A. Kasaeian, CFD modeling and predicting the performance of direct absorption of nanofluids in trough collector, Applied Thermal Engineering 148 (2019) 256–269.

[73] E.D. Reyes-Téllez, R.A. Conde-Gutiérrez, J.A. Perez, E. Cardoso, S. Silva Martínez, F. Sierra, et al., Optimal operating condition for a type parabolic trough collector with low-cost components using inverse neural network and solved by genetic algorithm, Desalination and Water Treatment 73 (2017) 80–89.

[74] T.E. Boukelia, O. Arslan, M.S. Mecibah, ANN-based optimization of a parabolic trough solar thermal power plant, Applied Thermal Engineering 107 (2016) 1210–1218.

[75] R. Loni, E.A. Asli-Ardeh, B. Ghobadian, M.H. Ahmadi, E. Bellos, GMDH modeling and experimental investigation of thermal performance enhancement of hemispherical cavity receiver using MWCNT/oil nanofluid, Solar Energy 171 (2018) 790–803.

[76] R. Loni, A. Kasaeian, K. Shahverdi, E. Askari Asli-Ardeh, B. Ghobadian, M.H. Ahmadi, ANN model to predict the performance of parabolic dish collector with tubular cavity receiver, Mechanics & Industry 18 (2017) 408.

[77] A. Khosravi, S. Syri, J.J.G. Pabon, O.R. Sandoval, B.C. Caetano, M.H. Barrientos, Energy modeling of a solar dish/Stirling by artificial intelligence approach, Energy Conversion and Management 199 (2019) 112021.

[78] Prediction of solar Stirling power generation in smart grid by GA-ANN model, International Journal of Computer Applications in Technology 55 (2017) 147–157.

[79] T.S. Vikram, K.S. Reddy, Investigation of convective and radiative heat losses from modified cavity based solar dish steam generator using ANN, International Journal of Thermal Sciences 87 (2015) 19–30.

[80] G. López, C.A. Gueymard, J.L. Bosch, I. Rapp-Arrarás, J. Alonso-Montesinos, I. Pulido-Calvo, et al., Modeling water vapor impacts on the solar irradiance reaching the receiver of a solar tower plant by means of artificial neural networks, Solar Energy 169 (2018) 34–39.

[81] A. Khosravi, M. Malekan, J.J.G. Pabon, X. Zhao, M.E.H. Assad, Design parameter modelling of solar power tower system using adaptive neuro-fuzzy inference system optimized with a combination of genetic algorithm and teaching learning-based optimization algorithm, Journal of Cleaner Production 244 (2020) 118904.

[82] W. Yaïci, M. Longo, E. Entchev, F. Foiadelli, Simulation study on the effect of reduced inputs of artificial neural networks on the predictive performance of the solar energy system, Sustainability 9 (2017) 1382.

[83] P. Schöttl, G. Bern, D.W. van Rooyen, J.A. Fernández Pretel, T. Fluri, P. Nitz, Optimization of Solar Tower molten salt cavity receivers for maximum yield based on annual performance assessment, Solar Energy 199 (2020) 278–294.

[84] G. Zhu, Development of an analytical optical method for linear Fresnel collectors, Solar Energy 94 (2013) 240–252.

[85] G. Zhu, T. Wendelin, M.J. Wagner, C. Kutscher, History, current state, and future of linear Fresnel concentrating solar collectors, Solar Energy 103 (2014) 639–652.

CHAPTER FOUR

Neural simulation of a solar thermal system in low temperature

Francisco J. Diez[1], Leticia Chico-Santamarta[2], Adriana Correa-Guimaraes[1], Andrés Martínez-Rodríguez[1] and Luis M. Navas-Gracia[1]
[1]University of Valladolid, Department of Agricultural and Forestry Engineering, University of Valladolid, Palencia, Castile and León, Spain
[2]Harper Adams University, Newport, Shropshire, United Kingdom

Contents

4.1 Introduction

Renewable energies are characterized as being infinite sources of energy at the human scale, not pollutant with their use (they contribute to a reduction of the greenhouse gas emissions associated with energy generation), distributed/decentralized (they contribute to the self-sufficiency and

Artificial Neural Networks for Renewable Energy Systems and Real-World Applications.
DOI: https://doi.org/10.1016/B978-0-12-820793-2.00005-7

energy independency of consumers and nations), they improve the energy efficiency of countries, and they boost business development and employment. Thus, renewable energies are considered sustainable sources of energy.

Amongst all the renewable energies, solar energy stands out from the others because it uses very flexible and scalable conversion technologies, which allows a great range of use, from electrical and thermal energy auto-generation applications in isolated installations, to generation power plans of large installed capacity connected to public distribution networks. Furthermore, solar energy can be used through the conversion of thermal energy or electrical energy to heat water at low temperatures [1], an application that is very well known and used for its technologic and economic viability, as the installation and use costs are clearly competitive for domestic use (residential buildings) and/or industrial and farming use. Thus Islam et al. [2] describe the different types of solar water heating and its market potential, concluding that these systems have a significant positive impact in the economic, environmental, and energy-efficiency perspectives of consumers, as a function of the commercial implementation degree.

A solar thermal system (STS) used for water heating can be described as an energy interface between environmental variables, which define the solar resource used, and variables for the calorific energy produced corresponding to generation, storage, and consumption. The STS, through a solar collector, absorbs incidental solar energy and transforms it into thermal energy, which is transported through a working heat transfer fluid to a heat exchanger, which is then transferred to the hot water used for consumption [3]. Also, low-temperature STSs can be active (open or closed loop) or passive (thermosiphon and integrated collector storage).

Solar collectors, storage tanks, and heat transfer fluids are the three main components of an STS, which have been reviewed in Ref. [4], identifying the lack of work done on the improvement of STS behavior, as it could be using nanofluids as heat transfer fluid. Shukla et al. [5] reviewed the design of solar water heating systems, the sizing of the components (i.e., solar collector, storage tank, heat exchanger, and heat transfer fluid), the actual technological developments to improve the performance, and the system's costs. Along these lines, Sadhishkumar and Balusamy [6] analyzed techniques to improve heating transfer between the primary circuit or collectors and the secondary or consumption circuit, amongst which are included the collector design, collector tilt angle, coating of pipes, fluid flow rate, thermal insulation, integrated collector storage, thermal energy storage, use of phase change materials, and insertion of twisted tapes.

On a different topic, Wang et al. [7] reviewed theoretically and from a practical point of view the technology available for solar water heating, analyzing its potential market and the challenges that still exist in the research topic. Models for the thermal simulation performance of STSs are divided by authors into two different groups: analytical models, to address the heat transfer and thermal balance, and transient energy models, to simulate the dynamic characteristics of the system. Jaisankar et al. [8] also performed a literature review of existing systems for solar water heating and the technological developments in which work is being done to improve the performance or to design new systems. Thirugnanasambandam et al. [9] studied the available STS technologies, analyzing their performance as a function of the typology and formulating simulation models of the components' behaviors.

Regarding STS modeling, Nogueira et al. [10] developed software to facilitate the design of small solar water heating systems using MATLAB® as a simulation platform, using the f-chart method applied in Brazil (Brazilian Technical Standards NBR-15569). Also, they reviewed the most commonly used computing tools to design solar water heating systems, such as: F-Chart software [11], developed by Klein and Beckman in the Solar Energy Laboratory at the University of Wisconsin, United States; German tool T*SOL [12], for designing and simulating; Polysun [13], used for simulation of active and passive STS; Canadian tool RETScreen International [14], which also applies the f-chart method; and others, such as TermoDim, Helios-Chart, Solterm, and SolarEnergy.

Along the same lines, the TRaNsient SYStem (TRNSYS) simulation software [15] is a very interesting STS simulation tool, achieving low prediction errors (from 5% to 10%). Shrivastava et al. [16] analyzed the application of STS software, including a comparative analysis with other modeling and simulation tools, such as EnergyPlus, RETScreen International, SolDesigner, SolarPro, T*SOL, WATSUN, F-Chart software, and Polysun. Abdunnabia et al. [17] studied the behavior of TRNSYS in the simulation of four evacuated tube solar collectors (type 538 model of TRNSYS), during six consecutive experimenting days and under Libyan meteorological conditions. The collectors were connected in series and a 300-L hot-water tank was used (type 38 model of TRNSYS), resulting in satisfactory differences with the daily averages (e.g., solar irradiation over the inclined plane, 3.0%–12.3%; solar collectors array temperature, 6.0%–19.3% in the inlet and 8.6%–23.1% in the outlet; accumulated energy 1.3%–31.6%; and hot water temperature lower than 16%).

Additionally, Ayompe et al. [18] presented a TRNSYS model validated for Irish conditions, which is used to simulate solar heating systems with forced water circulation, which are normally used in mild climates, with a hot water demand profile equivalent to a typical European domestic house. The results for the percentage mean absolute error found for the flat-plate solar collectors (type 73 model of TRNSYS) were: collector outlet temperature of 16.9%, collectors taken heat of 14.1%, and charged delivered heat of 6.9%. The model underestimated the temperature of the outlet fluid of the collector by 9.6%, and overestimated the generated heat by 7.6% and the charged delivered heat by 6.9%. Furthermore, the model was tested over a year [19], obtaining the following results: harnessed annual average daily energy = 19.6 MJ/d, supplied energy for the solar coil = 16.2 MJ/d, solar fraction = 32.2%, collector efficiency = 45.6%, system efficiency = 37.8%, and pipe supply losses = 16.4% of the captured energy.

In a similar vein, Khordehgah et al. [20] presented a TRNSYS model which was validated in London (UK). It simulates the behavior of a photovoltaic-thermal solar collector, to meet the domestic hot water demand of a house, through a connection between the solar thermal collector and a thermal storage tank, a bomb, and a controller. Antoniadis and Martinopoulos [21] worked with an integrated STS in a building, in which seasonal storage was optimized thanks to the use of TRNSYS. In the same way, Bava and Furbo [22] described a TRNSYS-MATLAB model to simulate the behavior of a large field of solar collector for urban heating applications in Denmark. In their model, they included aspects such as the flow distribution in different rows of the collector, the effect of the flow regime in the solar collector efficiency, the thermal capacity of the different components, and the effect of shadows amongst rows, making possible with this model the simulation of the real control strategy of a heating plant. In Mozambique [23], an hourly model was applied based on the energy balance of the STS components, comparing the obtained results with the simulation tool Polysun, concluding that the software tool does not achieve the correct sensitivity analysis for different STS parameters which affect the performance, given its wide applicability.

On the other hand, Lamnatou et al. [24,25] studied the advantages of the integration of solar technologies in buildings, as it is the case of building-integrated STSs, compared with building-added installations. In

Tunisia [26], the long-term performance of a thermosiphonic solar system was studied through a numeric model. It was demonstrated that the performance was greatly influenced by the tank and collector losses, which could decrease dramatically by using good insulation materials, such as polyurethane. Also, Hachicha et al. [27] reviewed models used to simulate and evaluate the optical, thermal, and dynamic performances of solar thermal collectors at high temperatures, such as the parabolic trough solar collector and solar tower. Yerdesh et al. [28] used numeric algorithms to predict the behavior of a solar collector combined with a heat bomb, which were integrated in a heating system in Kazakhstan. They used the Runge–Kutta method of fourth order, using Engineering Equation Solver software, to solve the equations of the heat conduction in the solar collector and the energy conservation in the working fluid. In Canada [29], the transitory modeling of the components of a heating solar system, the thermal charges, and its control, were studied using the simulation of the model in Simulink MATLAB and comparing the monthly average results with the RETScreen model. Dongellini et al. [30] performed an hourly simulation using Simulink MATLAB functional blocks that represent the STS components, taking advantage of the model to research the influence of the solar collector type, the tank volume, and the daily consumption profile.

Yaïci and Entchev [31] used artificial neural network (ANN) models to predict the performance of STS in Canada, using models for the prediction of the system's performance in applications with sanitary hot water and space heating, verifying the effectiveness of the models even when the input data were distorted by different levels of noise. Correa-Jullian et al. [32] built a physical model of a hot solar water plant using synthetic data of the TRNSYS software, predicting its performance with deep learning methods [e.g., ANN, recurrent neural network (RNN), and long short-term memory (LSTM)]. With these models the predictions obtained exceeded those obtained with other regression techniques, such as Bayesian ridge, Gaussian process, and linear regression. ANNs are a very useful technique to predict and optimize the performance of the different components of a solar energy system [33]. These authors reviewed, in an exhaustive way, the applications of this method, in addition to the ANN types and architectures commonly used and found in the literature.

In this study, a methodology is proposed for a simulation through ANNs for the operation of an active-type STS, with a closed loop for the heating of water at low temperature. ANN models are formulated for the main STS components (i.e., flat-plate solar collector, internal heat exchanger, and stratified tank working with dynamic and static modes), taking as a reference the data measured during the operation of an STS with the given characteristics located in the city of León (Castile and León, Spain). The short-term operation (1 minute) is modeled with the visual programming environment Simulink MATLAB. The developed neural simulation is implemented with different input data: solar irradiance (W/m^2), ambient temperature (°C), hot water outlet flow (L/min), and cold water inlet temperature (°C). In the work presented, different simulations are included with the purpose of proving the used methodology, comparing the results obtained with the measures taken at the experimental STS and the long-term predictions (over a period of 10 years), taking as a reference the data collected at the agrometeorological station, which belongs to SIAR network, located in Mansilla Mayor (Castile and León, Spain). The results were also compared with those taken from the f-chart method application using the CENSOLAR typical year.

4.2 Materials and methods

In this section, is the following are described: (1) meteorological database used to simulate the STS behavior; (2) the STS modeled and simulated; (3) ANN models obtained for each of the STS components; and (4) complete STS simulation model, implemented in MATLAB Simulink, with the use of ANN models for each previously obtained component.

4.2.1 Meteorological data

For the development of this work over a period of 10 years (2002 and 2004–12), data for horizontal hourly global solar irradiation were used as well as ambient temperature. The data were taken from the agrometeorological station located in Mansilla Mayor (Castile and León, Spain, Latitude: 42.512, longitude: 5.446; altitude: 791 MAMsl; local time: GMT—21.725555), which belongs to the SIAR network (Agroclimatic Information System for Irrigation, in Spanish) in Castile and León (Spain). The SIAR system is a project of the Environment and Rural Area, and Marine Ministry in Spain, which, in the region of Castile and León, is managed by the Agricultural Technological Institute of Castile and León (ITACyL), through the irrigation advisory service InfoRiego [34]. The information included in this database was processed to obtain the values of solar irradiance, with a 1-minute temporal scale, incidental over an inclined surface (on flat-plate solar collector), which were the data used for the neural simulation.

For that purpose, in order to obtain the value for the hourly global irradiation that is received over the inclined plane of the panels [35] calculations were done for hourly extraterrestrial solar irradiation [36–38], the solar height for each hour of the day [35], solar rays incidental angle over the inclined plane of the solar collectors [39], and estimation of the hourly diffuse solar irradiation over the horizontal [40].

Finally, the energetic value for the hourly global solar irradiation over the inclined plane of the collectors was converted to the power value of the corresponding hourly average solar irradiance. These hourly values were used, once they were interpolated with 1-minute intervals, to create the database used as input values of the neural simulation, with the objective of performing the most precise simulation of the STS operation over a period of 10 years [41]. Hourly data interpolation (solar irradiance and ambient temperature) to obtain its value each minute, was done using the MATLAB function Fcn: interp1((1:24)',u,(1:0.01598:24)') with an outlet dimension of 1440 (daily minutes). The input of this function is the hourly data matrix $[\text{day} \times \text{hour}]_{30 \times 24}$ and the output is the data matrix interpolated every minute during the whole month $[\text{day} \times \text{minute}]_{30 \times 1440}$.

4.2.2 Solar thermal system

A STS tested and simulated was located at the installations of the University of León (León, Castile and León, Spain; latitude: 42.614; longitude: −5.561). It was constituted as an active STS (forced circulation) which worked with a working fluid at low temperature (water with antifreeze). It was made with a flat-plate solar collector with a 1.85 m^2 surface absorber, formed with eight vertical interchangeable tubes welded to two horizontal collector tubes, including the group inside a case with a simple glass cover. The collector was installed with a 45° inclination in relation to the horizontal surface and south oriented. The STS also had a circulation bomb with three speeds (1–1.6–2 L/min) for the working fluid circulation in the collector's primary circuit, just like an expansion vessel, an automatic air purge, and a safety valve. The STS primary circuit was completed with an internal coil-type heat exchanger in a thermal water storage tank with a capacity of 150 L (Fig. 4.1). All the components were connected to each other through a thermally insulated copper pipe, and were assembled in a steel-profile structure with wheels to facilitate its transport.

The measured variables which are linked to the operation of the STS are the following:

Solar irradiance, G(t) (W/m^2), measured with a pyranometer located in the inclined plane of the collector. This is a first-class pyranometer, manufactured according to the ISO 9060 standard, following the recommendations given by the World Meteorological Organization, with a measurement range between 0 and 2000 W/m^2 and a spectral field between 305 and 2800 nm.

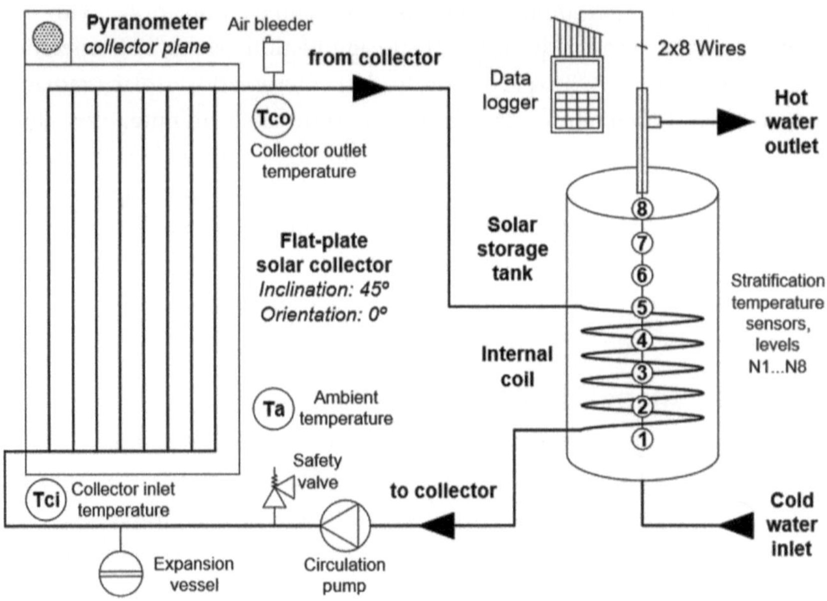

Figure 4.1 Connection diagram of the solar thermal system used in the work. Main components of the solar thermal system (flat-plate solar collector, storage tank, internal heat exchanger and circulation pump), safety elements (air bleed, safety valve and expansion vessel), connection (copper pipe), and measurement (pyranometer, temperature sensors and data logger). *Modified from F.J. Diez, Optimización del rendimiento en los sistemas solares térmicos mediante modelos neurales (in Spanish), University of Valladolid (2016). https://doi.org/10.35376/10324/21048.*

Ambient temperature, $T_a(t)$ (°C), measured with a Pt-100 platinum resistance, protected from direct insulation and wind.

Fluid temperature to solar collector, $T_{ci}(t)$ (°C), fluid temperature from solar collector, $T_{co}(t)$ (°C), and liquid fluid flow in the collector circuit, $C_c(t)$ (L/min), measured with each Pt-1000 sounding lines and a lineal displacement flowmeter, respectively.

Stratification temperatures of the storage tank, [i.e., $N_1(t)$, $N_2(t)$, $N_3(t)$, $N_4(t)$, $N_5(t)$, $N_6(t)$, $N_7(t)$, $N_8(t)$] (°C), measured with eight Pt-100 resistors, where the N_1 sensor was located at the bottom of the tank and the N_8 sensor was located at the top.

Sensors were connected either to a 16-bit and eight-measurement channel portable data logger, with the purpose of storing the stratification temperatures of the storage tank every minute, or to a data acquisition card in the computer, for the rest of the sensors. The card recorded information every 2 seconds, and saved the average data every minute.

4.2.3 Artificial neural networks of the components of the solar thermal system

Simulation of the STS behavior was done using the models based in ANNs for each of the components, which were previously formulated from the experimental data of its behavior, as described below.

Measured and registered data corresponding to the operation of each of the STS components were used to train and validate the formulated ANN models for each of them. The automatic pre- and postprocessing of the data was carried out using a MATLAB Neural Network Fitting Toolbox (nftool) Graphical User Interface (GUI) [42].

The creation of ANN models was done with the MATLAB function Fcn: (feedforwardnet), with the input and output data vectors determined by the size of the respective layers, generating multilayer feed-forward perceptron ANN models, with just one hidden layer, where the selected transfer function between the neurons was the hyperbolic sigmoidal tangent (tansig). On the other side, the selected transfer function for the outlet layer neuron was lineal (pureline). The back-propagation Levenberg–Marquardt algorithm was applied to achieve a quicker network optimization (trainlm), using the bias learning function and the weight moment with descendent gradient (learngdm). The ANN training was done with the MATLAB function Fcn: (train), for each ANN model, using the input data vector matrix and the outlet data vector matrix, recording the complete training process (epoch and performance). The

MATLAB function Fcn: (sim) was used with the previously trained ANNs, to perform the output simulation within the validation process, using the input data vector matrix different to the one used for the training, as in Ref. [43].

In the hidden layer of the ANN model, different neurons numbers were trained (i.e., 1, 2, 3, 4, 5, 6, 7, 8, 9, 10, 20, 30, 40, and 50), to select the neural network architecture with the best results in the outlet with respect to the measured data of the operation of the STS. The MATLAB GUI Neural Network Fitting Toolbox (nftool) showed the results for the training and simulation processes for the chosen ANN architecture, as a function of the chosen algorithms for the training. The selected architecture for each ANN model with the different input combinations, for the modeling of the flat-plate solar collector, internal heat exchanger and stratified tank, working with dynamic and static modes, is shown in Figs. 4.2–4.7, based on the following considerations:

The selected ANN model for the flat-plate solar collector [41, 44] has four input variables [G(t), T_a(t), T_{ci}(t), C_c(t)], with five neurons in the hidden layer, and one neuron in the outlet corresponding to the solar collector fluid temperature variable in the next time interval [T_{co}(t + 1)] in Fig. 4.2.

The selected ANN model for the internal heat exchanger [41], has 10 input variables [T_{co}(t), N_1(t), N_2(t), N_3(t), N_4(t), N_5(t), N_6(t), N_7(t), N_8(t), C_c(t)], with 10 neurons in the hidden layer, and one outlet neuron

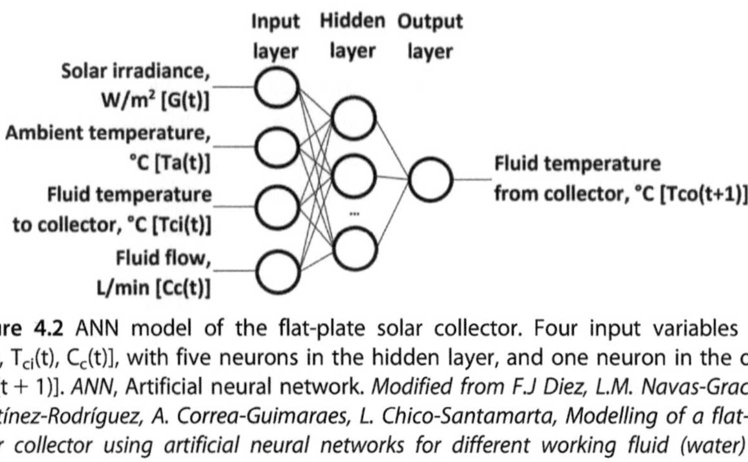

Figure 4.2 ANN model of the flat-plate solar collector. Four input variables [G(t), T_a(t), T_{ci}(t), C_c(t)], with five neurons in the hidden layer, and one neuron in the outlet [T_{co}(t + 1)]. *ANN*, Artificial neural network. *Modified from F.J Diez, L.M. Navas-Gracia, A. Martínez-Rodríguez, A. Correa-Guimaraes, L. Chico-Santamarta, Modelling of a flat-plate solar collector using artificial neural networks for different working fluid (water) flow rates, Solar Energy 188 (2019) 1320–1331. https://doi.org/10.1016/j.solener.2019.07.022.*

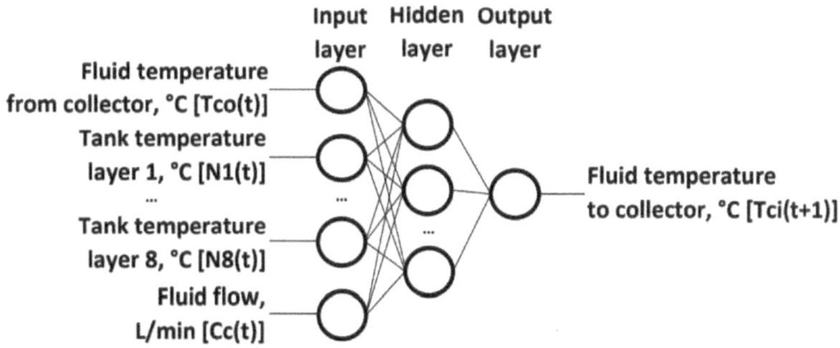

Figure 4.3 ANN model of the internal heat exchanger. Ten input variables [T_{co}(t), N_1(t), N_2(t), N_3(t), N_4(t), N_5(t), N_6(t), N_7(t), N_8(t), C_c(t)], with 10 neurons in the hidden layer, and one outlet neuron [T_{ci}(t + 1)]. *ANN*, Artificial neural network. *Modified from F.J. Diez, Optimización del rendimiento en los sistemas solares térmicos mediante modelos neurales (in Spanish), University of Valladolid (2016). https://doi.org/10.35376/10324/21048.*

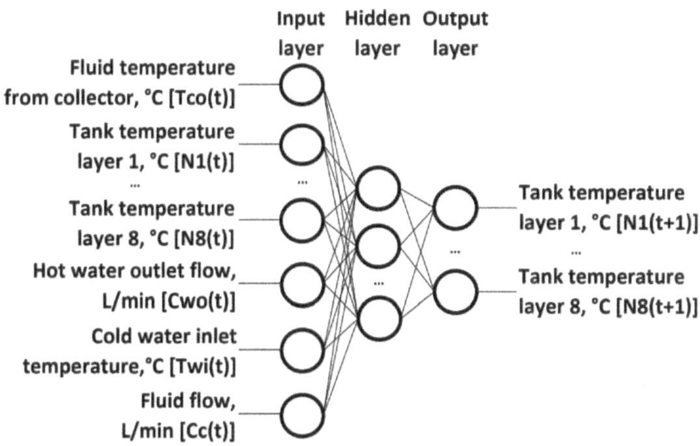

Figure 4.4 ANN model of the stratified tank (charge + discharge, dynamic mode). Twelve input variables [T_{co}(t), N_1(t), N_2(t), N_3(t), N_4(t), N_5(t), N_6(t), N_7(t), N_8(t), C_{wo}(t), T_{wi}(t), C_c(t)], with 22 neurons in the hidden layer, and eight outlet neurons [N_1(t + 1), N_2(t + 1), N_3(t + 1), N_4(t + 1), N_5(t + 1), N_6(t + 1), N_7(t + 1), N_8(t + 1)]. *ANN*, Artificial neural network. *Modified from F.J. Diez, Optimización del rendimiento en los sistemas solares térmicos mediante modelos neurales (in Spanish), University of Valladolid (2016). https://doi.org/10.35376/10324/21048.*

corresponding to the solar collector fluid temperature variable in the next time interval [T_{ci}(t + 1)] in Fig. 4.3.

The selected ANN model for the stratified tank, working with dynamic mode (charge + discharge) [41], has 12 inputs corresponding to the variables [T_{co}(t), N_1(t), N_2(t), N_3(t), N_4(t), N_5(t), N_6(t), N_7(t), N_8(t),

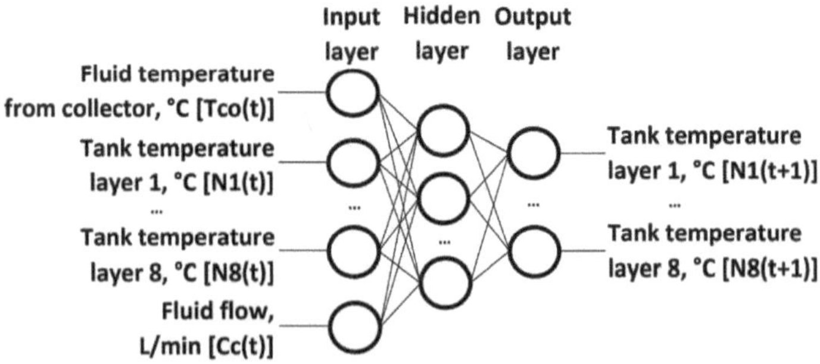

Figure 4.5 ANN model of the stratified tank (charge + no discharge, dynamic mode). Ten input variables [$T_{co}(t)$, $N_1(t)$, $N_2(t)$, $N_3(t)$, $N_4(t)$, $N_5(t)$, $N_6(t)$, $N_7(t)$, $N_8(t)$, $C_c(t)$], with 18 neurons in the hidden layer, and eight output neurons [$N_1(t+1)$, $N_2(t+1)$, $N_3(t+1)$, $N_4(t+1)$, $N_5(t+1)$, $N_6(t+1)$, $N_7(t+1)$, $N_8(t+1)$]. *ANN*, Artificial neural network. *Modified from F.J. Diez, Optimización del rendimiento en los sistemas solares térmicos mediante modelos neurales (in Spanish), University of Valladolid (2016). https://doi.org/10.35376/10324/21048.*

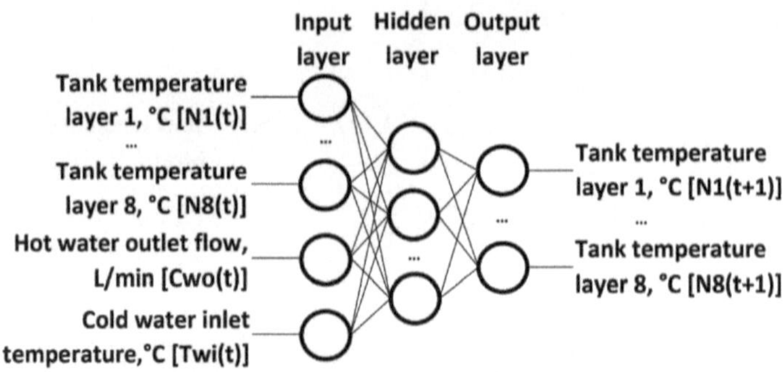

Figure 4.6 ANN model of the stratified tank (no charge + discharge, dynamic mode). Ten input variables [$N_1(t)$, $N_2(t)$, $N_3(t)$, $N_4(t)$, $N_5(t)$, $N_6(t)$, $N_7(t)$, $N_8(t)$, $C_{wo}(t)$, $T_{wi}(t)$], with 16 neurons in the hidden layer, and eight output neurons [$N_1(t+1)$, $N_2(t+1)$, $N_3(t+1)$, $N_4(t+1)$, $N_5(t+1)$, $N_6(t+1)$, $N_7(t+1)$, $N_8(t+1)$]. *ANN*, Artificial neural network. *Modified from F.J. Diez, Optimización del rendimiento en los sistemas solares térmicos mediante modelos neurales (in Spanish), University of Valladolid (2016). https://doi.org/10.35376/10324/21048.*

$C_{wo}(t)$, $T_{wi}(t)$, $C_c(t)$], with 22 neurons in the hidden layer, and eight outlet neurons for the tank temperatures variables in the next time interval [$N_1(t+1)$, $N_2(t+1)$, $N_3(t+1)$, $N_4(t+1)$, $N_5(t+1)$, $N_6(t+1)$, $N_7(t+1)$, $N_8(t+1)$] in Fig. 4.4.

The selected ANN model for the stratified tank, working with dynamic mode (charge + no discharge) [41], is structured into 10 inputs

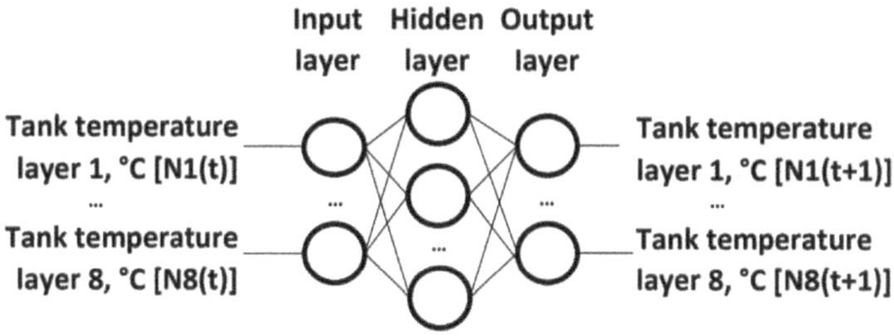

Figure 4.7 ANN model of the stratified tank (no charge + no discharge, static mode). Eight input variables [N_1(t), N_2(t), N_3(t), N_4(t), N_5(t), N_6(t), N_7(t), N_8(t)], with 16 neurons in the hidden layer, and eight output neurons [N_1(t + 1), N_2(t + 1), N_3(t + 1), N_4(t + 1), N_5(t + 1), N_6(t + 1), N_7(t + 1), N_8(t + 1)]. *ANN*, Artificial neural network. *Modified from F.J. Diez, L.M. Navas, A. Martínez, Prediction and identification of the evolution of solar-heated hot water temperature in a stratified tank in static mode using artificial neural networks. Applied Solar Energy (English Translation of Geliotekhnika), 52(3) (2016) 183–192. https://doi.org/10.3103/S0003701X16030051.*

referred to the variables [T_{co}(t), N_1(t), N_2(t), N_3(t), N_4(t), N_5(t), N_6(t), N_7(t), N_8(t), C_c(t)], with 18 neurons in the hidden layer, and eight output neurons for the tank temperature variables in the next interval of time [N_1(t + 1), N_2(t + 1), N_3(t + 1), N_4(t + 1), N_5(t + 1), N_6(t + 1), N_7(t + 1), N_8(t + 1)] in Fig. 4.5.

The selected ANN model for the stratified tank, working with dynamic mode (no charge + discharge) [41], consists of 10 input variables [N_1(t), N_2(t), N_3(t), N_4(t), N_5(t), N_6(t), N_7(t), N_8(t), C_{wo}(t), T_{wi}(t)], with 16 neurons in the hidden layer, and eight output neurons corresponding to the tank temperature variables in the next time interval [N_1(t + 1), N_2(t + 1), N_3(t + 1), N_4(t + 1), N_5(t + 1), N_6(t + 1), N_7(t + 1), N_8(t + 1)] in Fig. 4.6.

The formulated ANN model for the stratified tank, working with static mode (no charge + no discharge) [41, 45], has eight input for the variables [N_1(t), N_2(t), N_3(t), N_4(t), N_5(t), N_6(t), N_7(t), N_8(t)], with 16 neurons in the hidden layer, and eight output neurons for the tank temperature variables in the next time period [N_1(t + 1), N_2(t + 1), N_3(t + 1), N_4(t + 1), N_5(t + 1), N_6(t + 1), N_7(t + 1), N_8(t + 1)] in Fig. 4.7. In this case, since the temperature variations are much slower, the time interval between the selected ANN input/output was 10 minutes. Subsequently, an interpolation was performed to obtain the output in the time interval of one minute.

4.2.4 Neural simulation of the solar thermal system

ANN models for the STS components were converted into Simulink MATLAB blocks with a Fcn: (gensim) function, and with the blocks, a visual programming of the complete simulation model was performed in Fig. 4.8.

The behavioral simulations of the STS were performed with the following input data groups in the block (Inputs), from where the corresponding input blocks were distributed with a demultiplexer unit:

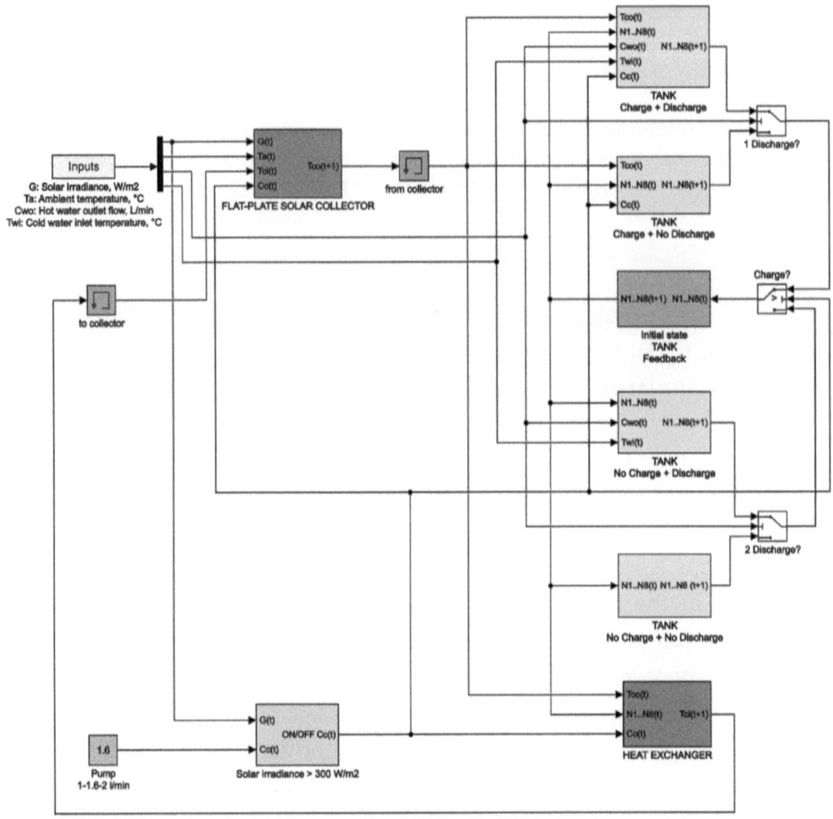

Figure 4.8 Simulation model of the solar thermal system, based on the ANN models of its components, for simulation in Simulink MATLAB. ANN model of the flat-plate solar collector of Fig. 4.2; ANN model of the internal heat exchanger of Fig. 4.3; and ANN models of the storage tank of (charge + discharge) in Fig. 4.4, (charge + no discharge) in Fig. 4.5, and (no charge + discharge) in Fig. 4.6 for the dynamic mode, and of (no charge + no discharge) in Fig. 4.7 for the static mode. *ANN*, Artificial neural network. *Modified from F.J. Diez, Optimización del rendimiento en los sistemas solares térmicos mediante modelos neurales (in Spanish), University of Valladolid (2016). https://doi.org/10.35376/10324/21048.*

- Solar irradiance, G(t) (W/m^2);
- Ambient temperature, $T_a(t)$ (°C);
- Cold water inlet temperature, $T_{wi}(t)$ (°C);
- Hot water outlet flow, $C_{wo}(t)$ (L/min).

Furthermore, delay units of the simulation of a time interval (i.e., 1 minute) were arranged between the blocks that represent the STS components. In effect, that is the way it was done in the outlet block (FLAT-PLATE SOLAR COLLECTOR) with a delay unit (from collector), and in the block outlet (HEAT EXCHANGER) with the delay unit (to collector). On the other side, the block (Initial state TANK Feedback) was made internally with eight delay units, one for each of the stratified temperatures for each of the tank layers [$N_1(t)$, $N_2(t)$, $N_3(t)$, $N_4(t)$, $N_5(t)$, $N_6(t)$, $N_7(t)$, $N_8(t)$]. There was also a demultiplexer situated in the inlet and a multiplexer at the outlet, which fed again the input values for the four blocks (TANK) and (HEAT EXCHANGER). These delay units were also used for the setting of the initial value of the variables at the beginning of the simulation.

The pump speed that propels the working fluid (three possible speeds 1, 1.6, or 2 L/min) was another value that was set before the simulation started through a constant (Pump). This speed set up the input of the comparison unit, and, as such, when the condition $G(t) > 300\ W/m^2$ was achieved, the pump was activated (change to ON) and if it was smaller the pump stopped (turns to OFF).

Finally, the design of the complete STS model in Simulink MATLAB had three switching units that were activated with the charge/discharge possible states, which depended on whether the circulation pump was operating or not and whether there was a discharge or not in the storage tank. In this way, according to the possible states of the switching units, the corresponding tank outlet (TANK) was forced to arrive (i.e., TANK Charge + Discharge, TANK Charge + No Discharge, TANK No Charge + Discharge, and TANK No Charge + No Discharge) to the block (Initial state TANK Feedback), which again fed the stratification temperatures values [i.e., $N_1(t+1)$,. . .,$N_8(t+1)$] toward the four block inlets (TANK) and for the block (HEAT EXCHANGER).

To carry out the neural simulation of the STS operation, a database with the inputs of the model was created:

1. Number of minutes per month;
2. Solar irradiance of the solar collector [G(t)];
3. Ambient temperature in the surroundings of the solar collector [$T_a(t)$];

4. Hot water outlet flow [$C_{wo}(t)$];
5. Cold water inlet temperature [$T_{wi}(t)$].

The stored energy in the tank, Q(Wh), was calculated using Eq. (4.1) [46]:

$$Q(\mathrm{Wh}) = m(\mathrm{kg})\cdot 1.16\left(\frac{\mathrm{Wh}}{\mathrm{kg}^\circ\mathrm{C}}\right)\cdot \Delta T(^\circ\mathrm{C}) \tag{4.1}$$

Taking into account the effect of the thermal stratification for the eight layers in which the tank was divided, Eq. (4.1) is transformed into Eq. (4.2):

$$Q(\mathrm{Wh}) = \sum_{i=1}^{8}\left(m_{\mathrm{N}i}(\mathrm{kg})\cdot 1.16\left(\frac{\mathrm{Wh}}{\mathrm{kg}^\circ\mathrm{C}}\right)\cdot(T_{\mathrm{N}i} - T_{\mathrm{Nw}i})(^\circ\mathrm{C})\right) \tag{4.2}$$

The mixed water temperature in the tank, T_{mix} (°C), can be calculated from the stored energy as in Eq. (4.3):

$$T_{mix}(^\circ\mathrm{C}) = \frac{Q(\mathrm{Wh})}{m(\mathrm{kg})\cdot 1.16\left(\frac{\mathrm{Wh}}{\mathrm{kg}^\circ\mathrm{C}}\right)} + T_{wi}(^\circ\mathrm{C}) \tag{4.3}$$

Or, based on the average of water temperatures in each of the tank layers, as in Eq. (4.4):

$$T_{mix}(^\circ\mathrm{C}) = \frac{\sum_{i=1}^{8}\left(m_{\mathrm{N}i}(\mathrm{kg})\cdot T_{\mathrm{N}i}(^\circ\mathrm{C})\right)}{\sum_{i=1}^{8} m_{\mathrm{N}i}(\mathrm{kg})} \tag{4.4}$$

With the layers in the tank being of the same volume, it can be considered that the water temperature in the tank which had been mixed homogeneously is shown in Eq. (4.5):

$$T_{mix}(^\circ\mathrm{C}) = \frac{\sum_{i=1}^{8} T_{\mathrm{N}i}(^\circ\mathrm{C})}{8} \tag{4.5}$$

where $T_{N1},\ldots,T_{N8}$ are the temperatures (°C) of the eight tank layers [N_1 (bottom),. . .,N_8 (top)].

With these calorific values, a hot water volume with the preparation temperature (i.e., 50°C in the current experiment) is achieved mixing the solar hot water with water from the supply network. With Eq. (4.6) the volume of hot water at the preparation temperature, which can be supplied by the thermal solar system to meet the consumption demands can be obtained:

$$V_{T_2} = \frac{T_1 - T_{cold}}{T_2 - T_{cold}} \cdot V_{T_1} \tag{4.6}$$

For the stratified tank, Eq. (4.6) is converted into Eq. (4.7):

$$V_{50°C} = \frac{T_{N8} - T_{wi}}{T_{50°C} - T_{wi}} \cdot V_{T_{N8}} \tag{4.7}$$

And for the mixed tank, it is transformed into Eq. (4.8):

$$V_{50°C} = \frac{T_{mix} - T_{wi}}{T_{50°C} - T_{wi}} \cdot V_{T_{mix}} \tag{4.8}$$

where $V_{50°C}$ is the hot water volume that is obtained by mixing the preparation mixture $T_{50°C}$ (i.e., 50°C in this research work), V_{TN8} is the volume of solar hot water in the output of the stratified tank, T_{mix} is the temperature of the solar hot water at the outlet of the mixing tank, and T_{wi} is the cold water temperature that enters of the network through the mixing valve.

The volume of hot water at the preparation temperature of 50°C will vary according to the solar irradiance used for the STS, because sometimes it will cover the volume demanded and at other times it will not be sufficient, which is the moment at which an energy auxiliary system will have to start working.

The performance of the STS is calculated with the percentage of the hot water demand supplied by the solar water stored in the tank at the preparation temperature, taking into account the effect of the stratification and for the case of using a mixed tank, in order to compare the subsequent effects.

4.3 Results

In this section the results are presented which were obtained with the neural simulation done with the complete STS model during one-day, one-year and 10-year trials, in addition to the f-chart method for comparison:

1. Daily trials (i.e., 10/10/2011 and 11/28/2011), with two different working fluid flows (1 and 1.6 L/min, respectively).
2. Trials done over a complete year (i.e., 2012), with a 2 L/min working fluid flow.
3. Trials done over a long period of time (i.e., 10 years, corresponding to the year 2002 and the years between 2004 and 2012). Results obtained are monthly, annual, and average performance of the studied STS, with a working fluid flow of 1.6 L/min, for the case of the stratified tank (i.e., eight layers, when the hottest water moves to the top of the tank) and the tank completely mixed (i.e., with uniform temperature, or unstratified).
4. Results obtained from the f-chart method application described by Beckman et al. [47] which represents the world known reference for the dimension of STSs at low temperature; it is also the reference used for this work.
5. Comparison of the STS performance prediction (evaluated by the solar fraction) that was obtained with its neural simulation (stratified and mixed tank) and with the f-chart method, during 10 years.

In order to carry out the validation and be able to compare the predictive model's behavior, the following statistical tools have been used: root mean square error (RMSE, °C); determination coefficient (R^2), model adjustment level indicator; Durbin–Watson (DW) coefficient, used to detect the first-order autocorrection between the data; mean percentage error (MPE), which allows the interpretation the prediction error bias; forecast accuracy (FA), used to make a measurement available for the precision of prediction models in the short term; and Akaike information criterion (AIC), used to select the most adequate model taking into account the number of used variables, penalizing the most complex models.

4.3.1 Neural simulation during a day

4.3.1.1 Neural simulation during the day 10/10/2011

The STS neural simulation performed on the 10/10/2011 (dynamic mode) and the night after (static mode) was done using the data vector model as an input [$G(t)$, $T_a(t)$, $C_{wo}(t)$, $T_{wi}(t)$], measured each minute in the STS, and the values are shown in Fig. 4.9, which include the information of a 100-L of hot water discharged, previously stored in the tank, with a $C_{wo} = 1.7$ L/min flow, that was done between 15h04 and 16h02.

Before the start of the simulation, tank stratified temperatures in the model were run to its initial state at 10h42 ($N_1 = 27.90°C$, $N_2 = 28.78°C$,

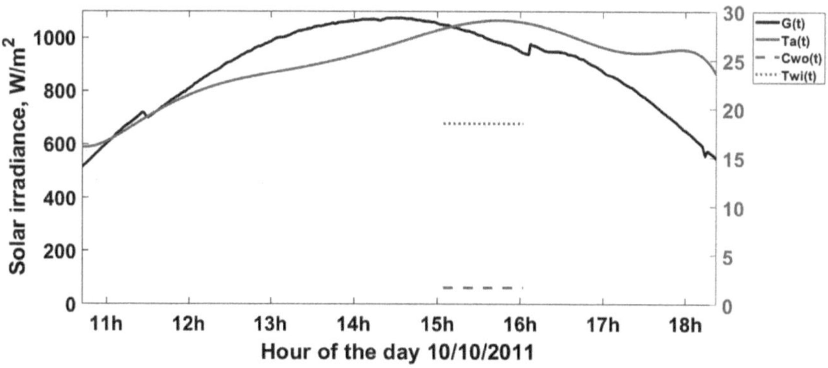

Figure 4.9 Input data for the neural simulation of the solar thermal system on 10/10/2011 (dynamic mode). G(t), solar irradiance (W/m^2); T_a(t), ambient temperature (°C); T_{wi}(t), cold water inlet temperature (°C); C_{wo}(t), hot water outlet flow (L/min). *Modified from F.J. Diez, Optimización del rendimiento en los sistemas solares térmicos mediante modelos neurales (in Spanish), University of Valladolid (2016). https://doi.org/10.35376/10324/21048.*

$N_3 = 29.17°C$, $N_4 = 29.52°C$, $N_5 = 29.56°C$, $N_6 = 29.62°C$, $N_7 = 29.34°C$, and $N_8 = 29.46°C$), specifically in the block shown in Fig. 4.8 (i.e., Initial state TANK Feedback). Also, the working speed of the circulation pump was fixed at 1 L/min [equivalent to a working fluid flow of 0.009 kg/(s m^2)] in the constant (Pump).

Daily simulation (dynamic mode) finished at 18h22 and the simulated results and STS measured data are shown in Fig. 4.10. The average statistical value for the RMSE was 0.77°C in the eight temperature layers of the stratified tank.

Night-time simulation (static mode) started at 18h30 and the STS model was adjusted, in the block (Initial state TANK Feedback, shown in Fig. 4.8) with the initial state of the tank stratified temperatures ($N_1 = 21.05°C$, $N_2 = 39.81°C$, $N_3 = 44.90°C$, $N_4 = 46.72°C$, $N_5 = 46.89°C$, $N_6 = 46.91°C$, $N_7 = 46.84°C$ and $N_8 = 49.10°C$).

Neural simulation was finished at 10h19 the day after. Results of the prediction, together with the STS measured data are shown in Fig. 4.11, where the average value for the RMSE is 0.13°C in the eight layers of the stratified temperatures tank.

Temperature statistical results for the eight tank layers are shown in Table 4.1 for the dynamic operation mode (daily period) and in Table 4.2 for the static mode of operation (night-time period).

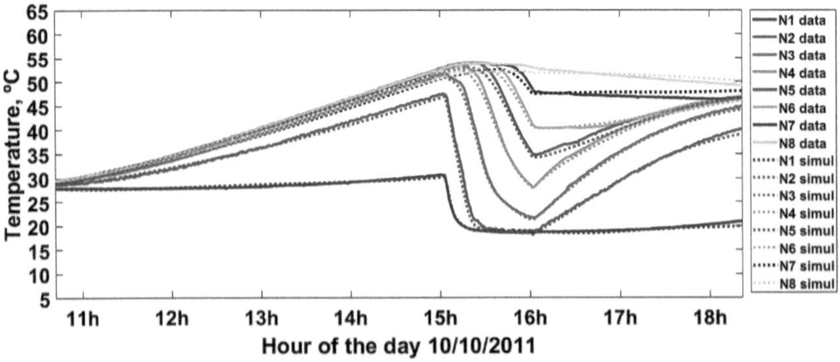

Figure 4.10 Data measured (data) and neural simulation (simul) every minute in the tank of the solar thermal system, during the dynamic mode on 10/10/2011, with a circulation pump speed of 1 L/min. Stratification temperatures in the eight layers of the tank, from N_1 (bottom) to N_8 (top). *Modified from F.J. Diez, Optimización del rendimiento en los sistemas solares térmicos mediante modelos neurales (in Spanish), University of Valladolid (2016). https://doi.org/10.35376/10324/21048.*

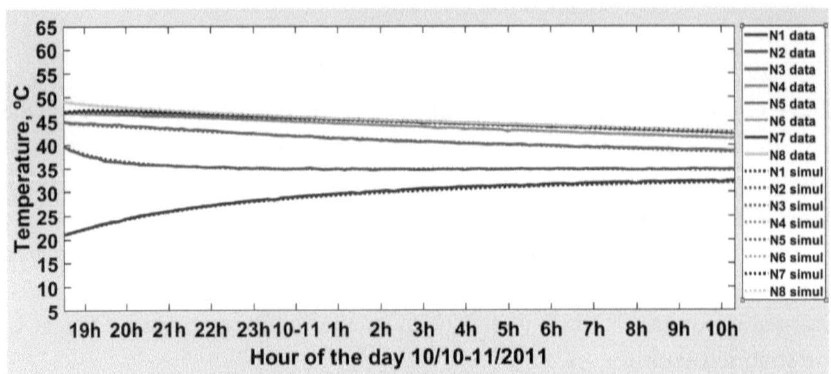

Figure 4.11 Data measured (data) and neural simulation (simul) every minute in the tank of the solar thermal system, during the static mode on 10/10–11/2011. Stratification temperatures in the eight layers of the tank, from N_1 (bottom) to N_8 (top). *Modified from F.J. Diez, L.M. Navas, A. Martínez, Prediction and identification of the evolution of solar-heated hot water temperature in a stratified tank in static mode using artificial neural networks. Applied Solar Energy (English Translation of Geliotekhnika), 52(3) (2016) 183–192. https://doi.org/10.3103/S0003701X16030051.*

4.3.1.2 Neural simulation during the day 11/28/2011

Input data for the STS neural simulation performed on the 11/28/2011 (dynamic mode) and the night of the day after (static mode) are shown in Fig. 4.12, which include the information about a discharge of 75 L of hot water in the tank, with a $C_{wo} = 1.75$ L/min flow that was performed

Table 4.1 Effectiveness of neural simulation vs data measured in the thermal stratification of the tank of the solar thermal system, during the dynamic mode on 10/10/2011, from 10h42 to 18h22.

Neural simulation vs tank temperature data (dynamic mode) 10h42–18h22, 10/10/2011

Tank layer	RMSE (°C)	R^2	DW	MPE	FA	AIC
N_1	0.3074	0.9956	0.0463	0.0028	0.9893	0.3128
N_2	0.6222	0.9935	0.1184	0.0094	0.9862	0.6330
N_3	0.4753	0.9962	0.0652	0.0057	0.9901	0.4836
N_4	0.5408	0.9940	0.0119	0.0082	0.9899	0.5503
N_5	0.7284	0.9884	0.0075	0.0127	0.9861	0.7411
N_6	0.8750	0.9831	0.0060	0.0131	0.9838	0.8903
N_7	1.1802	0.9731	0.0028	0.0089	0.9779	1.2008
N_8	1.0727	0.9818	0.0017	0.0133	0.9804	1.0915

AIC, Akaike information criterion; *DW*, Durbin–Watson; *FA*, forecast accuracy; *MPE*, mean percentage error; N_1, N_2, N_3, N_4, N_5, N_6, N_7, N_8, eight layers of the tank, from N_1 (bottom) to N_8 (top); *RMSE*, root mean square error (°C); R^2, determination coefficient.
Source: Modified from F.J. Diez, Optimización del rendimiento en los sistemas solares térmicos mediante modelos neurales (in Spanish), University of Valladolid (2016). https://doi.org/10.35376/10324/21048.

Table 4.2 Effectiveness of neural simulation vs data measured in the thermal stratification of the tank of the solar thermal system, during the static mode on 10/10–11/2011, from 18h30 to 10h19.

Neural simulation vs tank temperature data (static mode) 18h30–10h19, 10/10–11/2011

Tank layer	RMSE (°C)	R^2	DW	MPE	FA	AIC
N_1	0.2666	0.9917	0.0021	0.0081	0.9918	0.2711
N_2	0.1288	0.9775	0.0100	-0.0005	0.9972	0.1310
N_3	0.1548	0.9924	0.0064	0.0018	0.9968	0.1575
N_4	0.0539	0.9988	0.0289	0.0002	0.9990	0.0548
N_5	0.0736	0.9970	0.0151	0.0012	0.9985	0.0748
N_6	0.0723	0.9968	0.0133	0.0002	0.9986	0.0736
N_7	0.1091	0.9938	0.0070	-0.0015	0.9984	0.1110
N_8	0.0837	0.9975	0.0077	-0.0008	0.9985	0.0851

AIC, Akaike information criterion; *DW*, Durbin–Watson; *FA*, forecast accuracy; *MPE*, mean percentage error; N_1, N_2, N_3, N_4, N_5, N_6, N_7, N_8, eight layers of the tank, from N_1 (bottom) to N_8 (top); *RMSE*, root mean square error (°C); R^2, determination coefficient.
Source: Modified from F.J. Diez, Optimización del rendimiento en los sistemas solares térmicos mediante modelos neurales (in Spanish), University of Valladolid (2016). https://doi.org/10.35376/10324/21048.

between 15h32 and 16h14. These input data correspond with the data vector [G(t), T_a(t), C_{wo}(t), T_{wi}(t)] that was measured in the STS each minute.

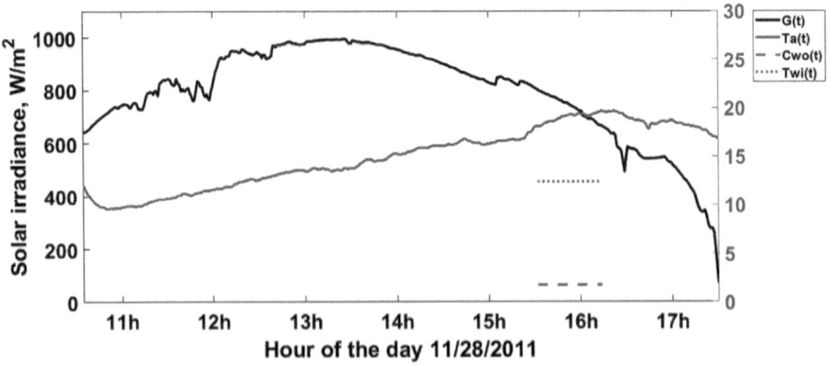

Figure 4.12 Input data to the neural simulation of the solar thermal system on 11/28/2011 (dynamic mode). G(t), solar irradiance (W/m^2); $T_a(t)$, ambient temperature (°C); $T_{wi}(t)$, cold water inlet temperature (°C); $C_{wo}(t)$, hot water outlet flow (L/min). *Modified from F.J. Diez, Optimización del rendimiento en los sistemas solares térmicos mediante modelos neurales (in Spanish), University of Valladolid (2016). https://doi.org/10.35376/10324/21048.*

Following the same procedure described in the previous section for the simulation done on 10/10/2011 and before the neural simulation started on 11/28/2011, the STS model block (Initial state TANK Feedback, shown in Fig. 4.8) was run to the initial state of the tank stratified temperatures at 10h32 ($N_1 = 23.67$°C, $N_2 = 24.48$°C, $N_3 = 25.23$°C, $N_4 = 25.62$°C, $N_5 = 25.70$°C, $N_6 = 25.82$°C, $N_7 = 25.70$°C, and $N_8 = 25.76$°C). Also, the working speed of the circulation pump 1.6 L/min [equivalent to a working fluid flow of 0.015 kg/(s m^2)] was run in the constant (Pump).

Daily simulation (dynamic mode) finished at 17h29, and the results are shown in Fig. 4.13 together with the STS measured data. The average value of the obtained RMSE for the eight tank stratification layers temperatures was 0.69°C.

On the other hand, the night-time neural simulation (static mode) started at 17h40 where the tank stratification temperatures were run ($N_1 = 13.28$°C, $N_2 = 23.88$°C, $N_3 = 29.10$°C, $N_4 = 31.68$°C, $N_5 = 32.78$°C, $N_6 = 40.45$°C, $N_7 = 45.56$°C, and $N_8 = 46.53$°C) in the block (Initial state TANK Feedback, shown in Fig. 4.8).

Neural simulation finished at 10h20 the day after, obtaining an RMSE average value of 0.41°C in the eight temperature layers of the stratification tank (results are shown in Fig. 4.14). In Fig. 4.14, the STS measured data are also presented.

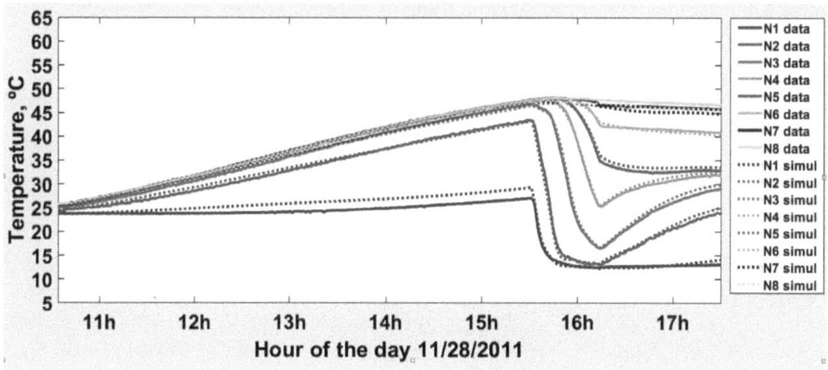

Figure 4.13 Data measured (data) and neural simulation (simul) every minute in the tank of the solar thermal system, during the dynamic mode on 11/28/2011, with a circulation pump speed of 1.6 L/min. Stratification temperatures in the eight layers of the tank, from N_1 (bottom) to N_8 (top). *Modified from F.J. Diez, Optimización del rendimiento en los sistemas solares térmicos mediante modelos neurales (in Spanish), University of Valladolid (2016). https://doi.org/10.35376/10324/21048.*

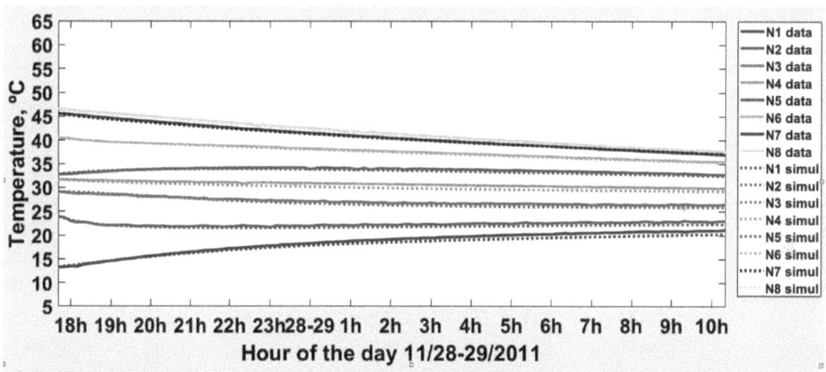

Figure 4.14 Data measured (data) and neural simulation (simul) every minute in the tank of the solar thermal system, during the static mode on 11/28–29/2011. Stratification temperatures in the eight layers of the tank, from N_1 (bottom) to N_8 (top). *Modified from F.J. Diez, L.M. Navas, A. Martínez, Prediction and identification of the evolution of solar-heated hot water temperature in a stratified tank in static mode using artificial neural networks. Applied Solar Energy (English Translation of Geliotekhnika), 52(3) (2016) 183–192. https://doi.org/10.3103/S0003701X16030051.*

Results of the statistical parameters RMSE, R^2, DW, MPE, FA, and AIC for the temperatures of the eight layers of the tank for the dynamic operation mode (daily period) are shown in Table 4.3. Table 4.4 shows the results for the static operation mode (night-time period).

Table 4.3 Effectiveness of neural simulation vs data measured in the thermal stratification of the tank of the solar thermal system, during the dynamic mode on 11/28/2011, from 10h32 to 17h29.

Neural simulation vs tank temperature data (dynamic mode) 10h32–17h29, 11/28/2011						
Tank layer	**RMSE (°C)**	**R^2**	**DW**	**MPE**	**FA**	**AIC**
N_1	1.3764	0.9307	0.0043	−0.0431	0.9517	1.4030
N_2	0.6177	0.9949	0.0756	−0.0110	0.9805	0.6296
N_3	0.5246	0.9960	0.0419	−0.0098	0.9848	0.5348
N_4	0.3627	0.9973	0.0258	−0.0044	0.9907	0.3697
N_5	0.4897	0.9946	0.0204	−0.0017	0.9900	0.4992
N_6	0.3965	0.9964	0.0236	0.0064	0.9917	0.4042
N_7	0.6237	0.9927	0.0062	0.0101	0.9871	0.6358
N_8	0.6096	0.9933	0.0080	0.0102	0.9878	0.6214

AIC, Akaike information criterion; *DW*, Durbin–Watson; *FA*, forecast accuracy; *MPE*, mean percentage error; N_1, N_2, N_3, N_4, N_5, N_6, N_7, N_8, eight layers of the tank, from N_1 (bottom) to N_8 (top); *RMSE*, root mean square error (°C); R^2, determination coefficient.
Source: Modified from F.J. Diez, Optimización del rendimiento en los sistemas solares térmicos mediante modelos neurales (in Spanish), University of Valladolid (2016). https://doi.org/10.35376/10324/21048.

Table 4.4 Effectiveness of neural simulation vs data measured in the thermal stratification of the tank of the solar thermal system, during the static mode on 11/28–29/2011, from 17h40 to 10h20.

Neural simulation vs tank temperature data (static mode) 17h40–10h20, 11/28–29/2011						
Tank layer	**RMSE (°C)**	**R^2**	**DW**	**MPE**	**FA**	**AIC**
N_1	0.5935	0.9260	0.0001	0.0256	0.9729	0.6031
N_2	0.4559	0.8603	0.0007	0.0183	0.9811	0.4632
N_3	0.4178	0.8805	0.0004	0.0125	0.9857	0.4246
N_4	0.6841	0.9314	0.0000	0.0214	0.9785	0.6951
N_5	0.2850	0.8411	0.0008	0.0068	0.9922	0.2896
N_6	0.1961	0.9795	0.0003	0.0049	0.9950	0.1993
N_7	0.1455	0.9964	0.0006	0.0025	0.9970	0.1479
N_8	0.1675	0.9958	0.0015	0.0033	0.9965	0.1702

AIC, Akaike information criterion; *DW*, Durbin–Watson; *FA*, forecast accuracy; *MPE*, mean percentage error; N_1, N_2, N_3, N_4, N_5, N_6, N_7, N_8, eight layers of the tank, from N_1 (bottom) to N_8 (top); *RMSE*, root mean square error (°C); R^2, determination coefficient.
Source: Modified from F.J. Diez, Optimización del rendimiento en los sistemas solares térmicos mediante modelos neurales (in Spanish), University of Valladolid (2016). https://doi.org/10.35376/10324/21048.

4.3.2 Neural simulation during 2012

The STS complete model results of the simulation over a year (i.e., 2012) which were done with Simulink MATLAB, were obtained using the input data vector [G(t), T_a(t), C_{wo}(t), T_{wi}(t)] for each minute. Solar irradiance data, G(t), coming from the SIAR database system, were previously used to obtain the solar irradiance value over the inclined collector surface. Hot water outlet flow, C_{wo}(t), values used for the simulation were obtained using the standard value of 75 L/day. This value and the normalized profile of the hourly distribution of the domestic consumption can be found in the topical literature ([48], and [37], p. 410). Similarly, the cold water inlet temperature variable, T_{wi}(t), was estimated from the network water measurements done during each month in 2011 which was provided to the tank.

Before the start of the neural prediction, and following the general procedure of the simulation developed in this work, the tank stratification temperatures at 00h00 on the first day of each month of the year [N_1(t), N_2(t), N_3(t), N_4(t), N_5(t), N_6(t), N_7(t), N_8(t)] were run in the STS model block (Initial state TANK Feedback, shown in Fig. 4.8) implemented in Simulink MATLAB. Also, the working speed of the circulation pump 2 L/min equivalent to a working fluid flow of 0.018 kg/(s m^2) was run in the constant (Pump). The simulation finished at 23h59 on the last day of each month, obtaining results of the simulations for the eight temperature layers in which it is considered that the tank is stratified, for each month in 2012, as is shown for January in Fig. 4.15, for February in Fig. 4.16, for March in Fig. 4.17, for April in Fig. 4.18, for May in Fig. 4.19, for June in Fig. 4.20, for July in Fig. 4.21, for August in Fig. 4.22, for September in Fig. 4.23, for October in Fig. 4.24, for November in Fig. 4.25, and for December in Fig. 4.26.

4.3.3 Neural simulation for 10 years

The simulation of the operation of the STS model implemented in Simulink MATLAB was also done with data obtained over a long period of time, specifically 10 years (2002 and 2004–2012); data that correspond to the SIAR system database. STS monthly and annual performance results (evaluated through the solar fraction), with a working speed of the circulation pump of 1.6 L/min [equivalent to a working fluid flow of 0.015 kg/(s m^2)] and considering the effect of the water thermal stratification inside the storage tank, are shown in Table 4.5. Table 4.6 includes the results referred to the tank with a homogeneous temperature inside.

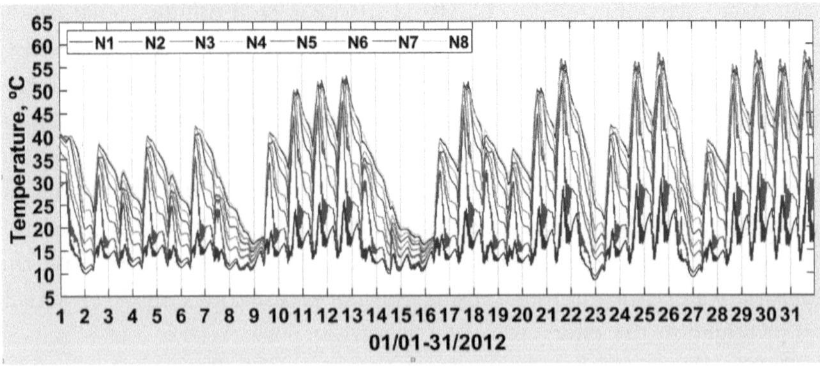

Figure 4.15 Neural simulation every minute during the month of January 2012 in the tank of the solar thermal system, with a circulation pump speed of 2 L/min. Stratification temperatures in the eight layers of the tank, from N_1 (bottom) to N_8 (top). *Modified from F.J. Diez, Optimización del rendimiento en los sistemas solares térmicos mediante modelos neurales (in Spanish), University of Valladolid (2016). https://doi.org/10.35376/10324/21048.*

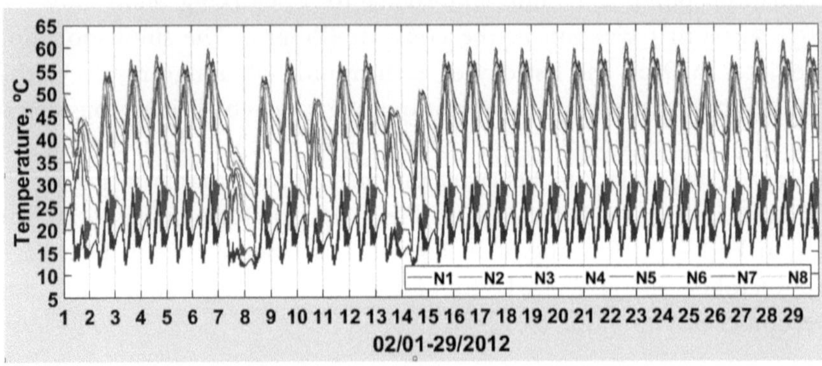

Figure 4.16 Neural simulation every minute during the month of February 2012 in the tank of the solar thermal system, with a circulation pump speed of 2 L/min. Stratification temperatures in the eight layers of the tank, from N_1 (bottom) to N_8 (top). *Modified from F.J. Diez, Optimización del rendimiento en los sistemas solares térmicos mediante modelos neurales (in Spanish), University of Valladolid (2016). https://doi.org/10.35376/10324/21048.*

4.3.4 f-Chart method

STS monthly and annual performance results (solar fraction) applying the f-chart method methodology in the established conditions for its application included in Ref. [47], are shown in Table 4.7. The results are obtained with the tested flat-plate solar collector with a working fluid

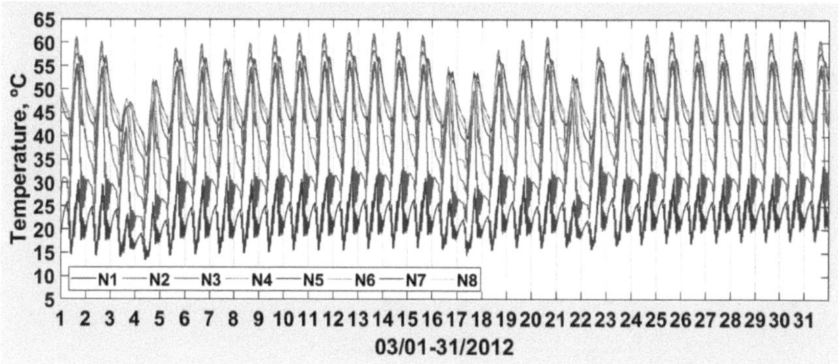

Figure 4.17 Neural simulation every minute during the month of March 2012 in the tank of the solar thermal system, with a circulation pump speed of 2 L/min. Stratification temperatures in the eight layers of the tank, from N_1 (bottom) to N_8 (top). *Modified from F.J. Diez, Optimización del rendimiento en los sistemas solares térmicos mediante modelos neurales (in Spanish), University of Valladolid (2016). https://doi.org/10.35376/10324/21048.*

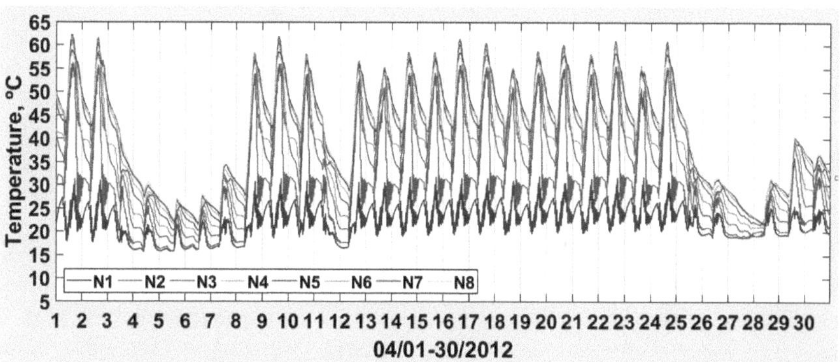

Figure 4.18 Neural simulation every minute during the month of April 2012 in the tank of the solar thermal system, with a circulation pump speed of 2 L/min. Stratification temperatures in the eight layers of the tank, from N_1 (bottom) to N_8 (top). *Modified from F.J. Diez, Optimización del rendimiento en los sistemas solares térmicos mediante modelos neurales (in Spanish), University of Valladolid (2016). https://doi.org/10.35376/10324/21048.*

flow of 1.6 L/min [0.015 kg/(s m^2)] and a monthly average daily solar radiation in the studied area for a typical CENSOLAR year [49].

C_w, hot water required for a day (L/d); M, number of days in a month (d); T_w, minimum acceptable hot water temperature (°C); T_m, mains water supply temperature (°C); L, monthly total heating load (MJ); A_c, flat-plate solar collector area (m^2); F_R, solar collector heat extraction

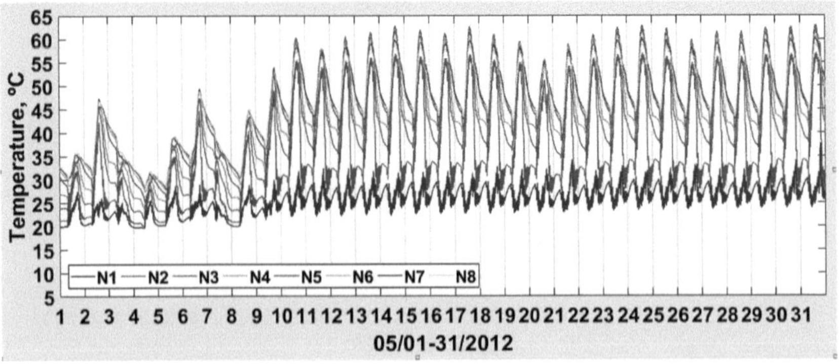

Figure 4.19 Neural simulation every minute during the month of May 2012 in the tank of the solar thermal system, with a circulation pump speed of 2 L/min. Stratification temperatures in the eight layers of the tank, from N_1 (bottom) to N_8 (top). *Modified from F.J. Diez, Optimización del rendimiento en los sistemas solares térmicos mediante modelos neurales (in Spanish), University of Valladolid (2016). https://doi.org/10.35376/10324/21048.*

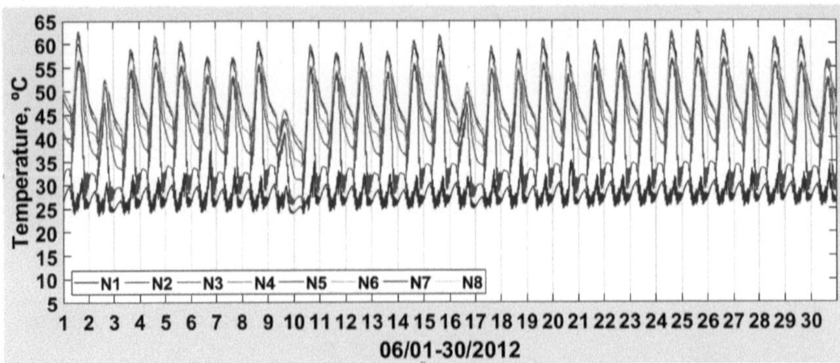

Figure 4.20 Neural simulation every minute during the month of June 2012 in the tank of the solar thermal system, with a circulation pump speed of 2 L/min. Stratification temperatures in the eight layers of the tank, from N_1 (bottom) to N_8 (top). *Modified from F.J. Diez, Optimización del rendimiento en los sistemas solares térmicos mediante modelos neurales (in Spanish), University of Valladolid (2016). https://doi.org/10.35376/10324/21048.*

efficiency factor (dimensionless); $(\tau\alpha)_n$, effective transmittance-absorptance product at normal incidence (dimensionless); F_R', solar collector-heat exchanger efficiency factor (dimensionless); H, monthly average daily radiation on a horizontal surface (MJ/m^2); β, inclination of a surface from the horizontal position (degrees); γ, surface azimuth angle, that is, the deviation of the normal to the surface with respect to the local meridian

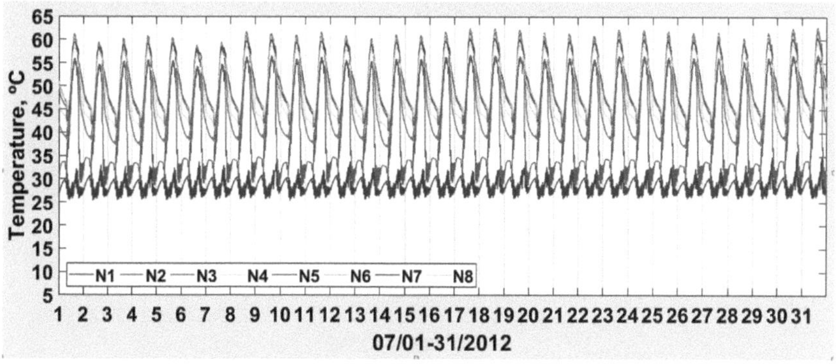

Figure 4.21 Neural simulation every minute during the month of July 2012 in the tank of the solar thermal system, with a circulation pump speed of 2 L/min. Stratification temperatures in the eight layers of the tank, from N_1 (bottom) to N_8 (top). *Modified from F.J. Diez, Optimización del rendimiento en los sistemas solares térmicos mediante modelos neurales (in Spanish), University of Valladolid (2016). https://doi.org/10.35376/10324/21048.*

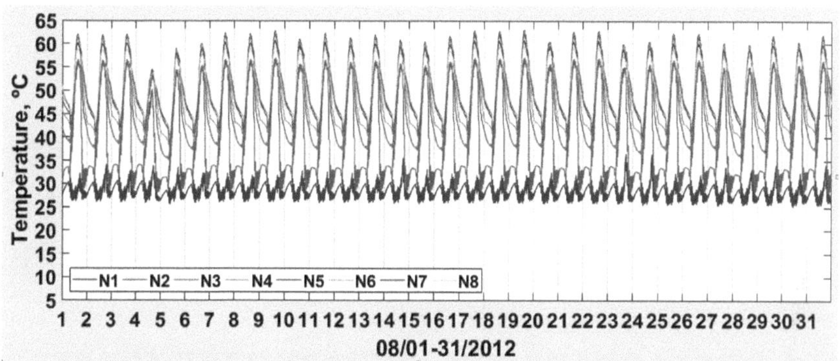

Figure 4.22 Neural simulation every minute during the month of August 2012 in the tank of the solar thermal system, with a circulation pump speed of 2 L/min. Stratification temperatures in the eight layers of the tank, from N_1 (bottom) to N_8 (top). *Modified from F.J. Diez, Optimización del rendimiento en los sistemas solares térmicos mediante modelos neurales (in Spanish), University of Valladolid (2016). https://doi.org/10.35376/10324/21048.*

(south zero, east positive) (degrees); R, ratio of the monthly average daily radiation on a tilted surface to that on a horizontal surface (dimensionless); Y, ratio of the total energy absorbed on the collector plate surface to the total heating load during the month (dimensionless); U_L, solar collector overall energy loss coefficient [W/(°C m^2)]; T_a, ambient temperature (°C); K_1, storage size correction factor; K_2, domestic water heating

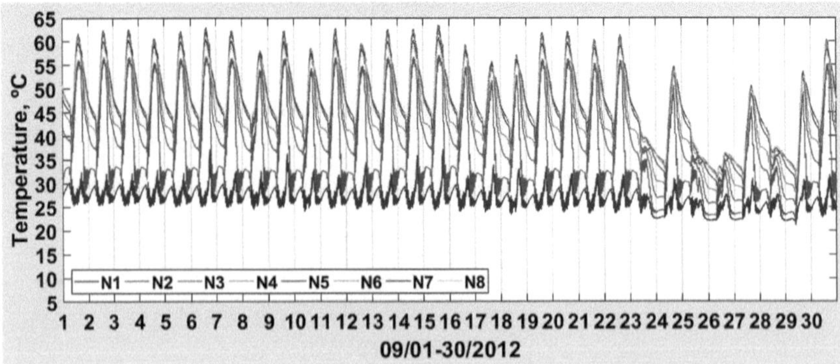

Figure 4.23 Neural simulation every minute during the month of September 2012 in the tank of the solar thermal system, with a circulation pump speed of 2 L/min. Stratification temperatures in the eight layers of the tank, from N_1 (bottom) to N_8 (top). *Modified from F.J. Diez, Optimización del rendimiento en los sistemas solares térmicos mediante modelos neurales (in Spanish), University of Valladolid (2016). https://doi.org/10.35376/10324/21048.*

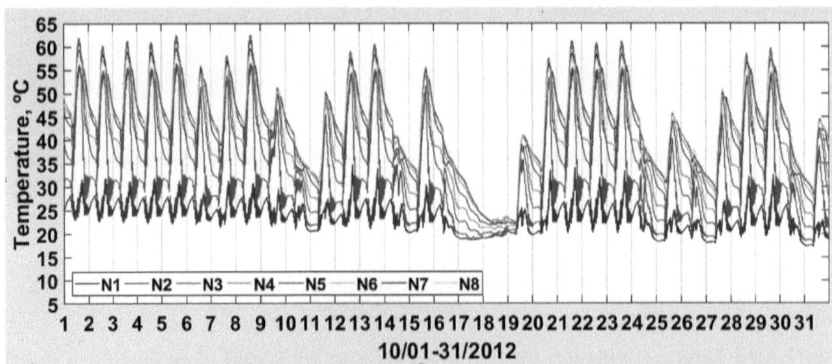

Figure 4.24 Neural simulation every minute during the month of October 2012 in the tank of the solar thermal system, with a circulation pump speed of 2 L/min. Stratification temperatures in the eight layers of the tank, from N_1 (bottom) to N_8 (top). *Modified from F.J. Diez, Optimización del rendimiento en los sistemas solares térmicos mediante modelos neurales (in Spanish), University of Valladolid (2016). https://doi.org/10.35376/10324/21048.*

correction factor; X, ratio of a reference collector energy loss to the total heating load during the month (dimensionless); f, fraction of the heating load supplied by solar energy for a month [0−1]; and F, fraction of the annual heating load supplied by solar energy (%).

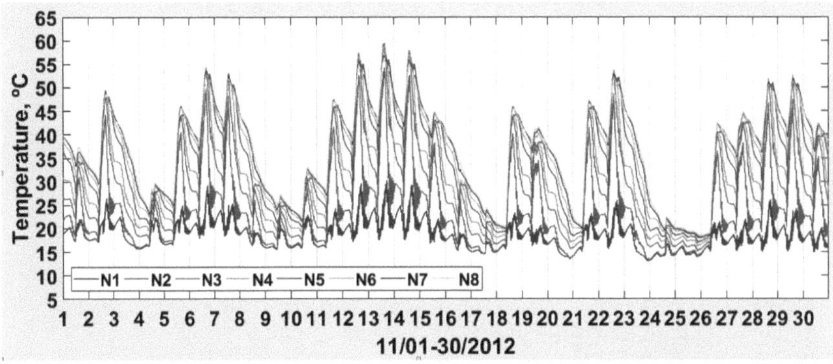

Figure 4.25 Neural simulation every minute during the month of November 2012 in the tank of the solar thermal system, with a circulation pump speed of 2 L/min. Stratification temperatures in the eight layers of the tank, from N_1 (bottom) to N_8 (top). *Modified from F.J. Diez, Optimización del rendimiento en los sistemas solares térmicos mediante modelos neurales (in Spanish), University of Valladolid (2016). https://doi.org/10.35376/10324/21048.*

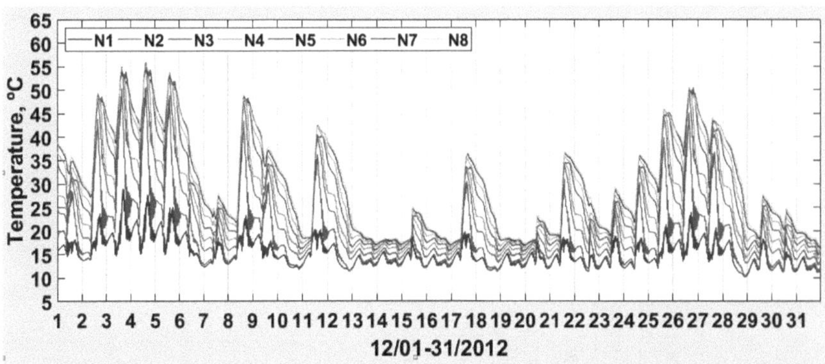

Figure 4.26 Neural simulation every minute during the month of December 2012 in the tank of the solar thermal system, with a circulation pump speed of 2 L/min. Stratification temperatures in the eight layers of the tank, from N_1 (bottom) to N_8 (top). *Modified from F.J. Diez, Optimización del rendimiento en los sistemas solares térmicos mediante modelos neurales (in Spanish), University of Valladolid (2016). https://doi.org/10.35376/10324/21048.*

4.3.5 Neural simulation versus f-chart method

The results are shown in Fig. 4.27 with a comparison between the STS annual performance results obtained over a period of 10 years (2002 and 2004–12), calculated with the STS complete model of the neural simulation in Simulink MATLAB, with the hot water storage tank thermal

Table 4.5 Annual and monthly performance (solar fraction) of the neural simulation of the solar thermal system with the stratified tank and a circulation pump speed of 1.6 L/min. Annual, annual performance (%); Jan, Feb,. . ., Dec, monthly performance (%).

Annual and monthly performance of the neural simulation with a stratified tank

Year	Annual	Jan	Feb	Mar	Apr	May	Jun	Jul	Aug	Sep	Oct	Nov	Dec
2002	76.74	46.45	77.06	78.71	89.59	88.82	93.75	100.24	96.36	87.79	71.96	48.15	42.44
2004	84.27	61.02	81.16	81.74	93.61	90.19	102.59	99.65	93.24	100.52	72.94	64.37	68.14
2005	86.46	62.55	88.90	92.05	85.11	94.42	103.74	103.88	103.52	99.58	68.80	67.03	68.60
2006	83.11	68.15	86.09	78.20	95.87	99.70	99.60	102.80	97.53	89.06	69.07	51.61	60.08
2007	85.50	48.32	68.12	88.09	91.23	91.91	94.75	102.68	100.64	97.93	90.18	88.96	62.56
2008	85.44	60.46	79.79	96.43	85.08	85.33	99.72	103.39	102.80	95.00	87.10	63.87	63.36
2009	85.29	46.94	85.48	97.52	92.44	98.72	96.04	103.45	102.25	102.63	86.78	59.74	51.90
2010	82.39	54.70	68.38	79.34	92.87	90.76	91.21	106.33	104.45	93.23	83.82	73.96	48.98
2011	84.14	46.45	79.26	79.39	92.72	99.38	102.22	102.19	98.81	100.98	93.55	55.25	59.55
2012	85.78	69.49	96.98	100.87	76.42	90.55	100.55	104.80	103.62	93.51	80.76	63.35	49.28
Mean	83.91	56.45	81.12	87.23	89.49	92.98	98.42	102.94	100.32	96.02	80.50	63.63	57.49

Source: Modified from F.J. Diez, Optimización del rendimiento en los sistemas solares térmicos mediante modelos neurales (in Spanish), University of Valladolid (2016). https://doi.org/10.35376/10324/21048.

Table 4.6 Annual and monthly performance (solar fraction) of the neural simulation of the solar thermal system with the mixed tank and a circulation pump speed of 1.6 L/min. Annual, annual performance (%); Jan, Feb,. . ., Dec, monthly performance (%).

Annual and monthly performance of the neural simulation with a mixed tank

Year	Annual	Jan	Feb	Mar	Apr	May	Jun	Jul	Aug	Sep	Oct	Nov	Dec
2002	60.75	31.81	58.55	61.99	72.81	72.42	77.66	83.58	79.83	71.65	55.66	33.88	29.45
2004	67.35	43.56	62.50	64.83	76.73	74.03	86.10	82.90	77.18	83.59	56.69	48.53	50.00
2005	69.86	46.93	70.18	73.96	68.41	77.98	87.30	87.22	86.66	82.95	56.08	50.30	50.72
2006	66.56	50.33	67.17	60.84	78.90	83.22	83.50	86.15	81.20	73.06	53.98	37.52	43.14
2007	68.76	33.72	49.97	70.34	74.19	75.67	78.65	86.20	83.83	81.49	73.66	71.04	45.63
2008	68.44	43.49	60.58	78.93	68.64	69.10	83.24	86.59	86.11	78.37	69.54	47.94	46.20
2009	68.86	32.14	67.88	80.09	75.26	82.37	80.08	86.85	85.56	85.79	70.05	43.97	36.50
2010	65.86	39.81	50.11	61.71	75.92	74.81	75.90	89.75	87.85	76.69	66.82	55.61	34.56
2011	67.84	34.25	60.53	62.30	76.23	82.96	85.88	85.82	82.25	84.16	76.69	40.12	42.70
2012	69.05	50.88	77.43	83.40	60.91	74.98	84.00	88.23	86.82	77.07	64.55	46.43	34.51
Mean	67.33	40.69	62.49	69.84	72.80	76.75	82.23	86.33	83.73	79.48	64.37	47.53	41.34

Source: Modified from F.J. Diez, Optimización del rendimiento en los sistemas solares térmicos mediante modelos neurales (in Spanish), University of Valladolid (2016). https://doi.org/10.35376/10324/21048.

Table 4.7 Calculation of the annual and monthly performance (solar fraction) of the solar thermal system applying the f-chart method. Meteorological data (monthly average daily radiation on a horizontal surface and ambient temperature), typical year CENSOLAR (Center for Solar Energy Studies, in Spanish); circulation pump speed of 1.6 L/min.

Annual and monthly performance with the f-chart method for the typical year (CENSOLAR)

Month	Jan	Feb	Mar	Apr	May	Jun	Jul	Aug	Sep	Oct	Nov	Dec
C_w	75	75	75	75	75	75	75	75	75	75	75	75
M	31	28	31	30	31	30	31	31	30	31	30	31
T_w	50	50	50	50	50	50	50	50	50	50	50	50
T_m	4	5	7	9	10	11	12	11	10	9	7	4
L	447.79	395.67	418.59	386.25	389.39	367.40	369.92	379.65	376.83	399.12	405.09	447.79
A_c	1.85	1.85	1.85	1.85	1.85	1.85	1.85	1.85	1.85	1.85	1.85	1.85
$F_R(\tau\alpha)_n$	0.46	0.46	0.46	0.46	0.46	0.46	0.46	0.46	0.46	0.46	0.46	0.46
$\frac{(\tau\alpha)}{(\tau\alpha)_n}$	0.96	0.96	0.96	0.96	0.96	0.96	0.96	0.96	0.96	0.96	0.96	0.96
$\frac{F_R'}{F_R}$	0.95	0.95	0.95	0.95	0.95	0.95	0.95	0.95	0.95	0.95	0.95	0.95
$H_{\beta\gamma}$	8.29	11.48	16.28	17.88	18.33	19.89	22.74	21.94	21.15	14.87	10.99	7.39
H	5.80	8.70	13.80	17.20	19.50	22.10	24.20	20.90	17.20	10.40	7.00	4.80
R	1.43	1.32	1.18	1.04	0.94	0.9	0.94	1.05	1.23	1.43	1.57	1.54
Y L	199.54	249.56	391.78	416.49	441.01	463.10	547.30	527.98	492.58	357.81	255.88	177.84
F_RU_L	2.3	2.3	2.3	2.3	2.3	2.3	2.3	2.3	2.3	2.3	2.3	2.3

T_a	5	6	10	12	15	19	22	22	19	14	9	6
K_1	0.98	0.98	0.98	0.98	0.98	0.98	0.98	0.98	0.98	0.98	0.98	0.98
K_2	0.78	0.80	0.82	0.88	0.87	0.85	0.84	0.79	0.80	0.84	0.84	0.76
X L	790.38	728.66	790.17	796.33	789.96	708.78	699.49	658.51	669.12	773.61	788.52	765.75
Y	0.44	0.63	0.93	1.07	1.13	1.26	1.47	1.39	1.30	0.89	0.63	0.39
X	1.76	1.84	1.88	2.06	2.02	1.92	1.89	1.73	1.77	1.93	1.94	1.71
f	0.32	0.46	0.67	0.75	0.78	0.85	0.96	0.92	0.88	0.64	0.46	0.28
F							65.48					

Cw, hot water required for a day (lL/d); *M*, number of days in a month (d); T_w, minimum acceptable hot water temperature (°C); T_m, mains water supply temperature (°C); *L*, monthly total heating load (MJ); A_c, flat-plate solar collector area (m^2); F_R, solar collector heat extraction efficiency factor (dimensionless); $(\tau\alpha)_n$, effective transmittance-absorptance product at normal incidence (dimensionless); F_R', solar collector-heat exchanger efficiency factor (dimensionless); *H*, monthly average daily radiation on a horizontal surface (MJ/m^2); β, inclination of a surface from the horizontal position (degrees); γ, surface azimuth angle, that is, the deviation of the normal to the surface with respect to the local meridian (south zero, east positive) (degrees); *R*, ratio of the monthly average daily radiation on a tilted surface to that on a horizontal surface (dimensionless); *Y*, ratio of the total energy absorbed on the collector plate surface to the total heating load during the month (dimensionless); U_L, solar collector overall energy loss coefficient [W/(°C m^2)]; T_a, ambient temperature (°C); K_1, storage size correction factor; K_2, domestic water heating correction factor; *X*, ratio of a reference collector energy loss to the total heating load during the month (dimensionless); *f*, fraction of the heating load supplied by solar energy for a month [0–1]; and *F*, fraction of the annual heating load supplied by solar energy (%)
Source: Modified from F.J. Diez, Optimización del rendimiento en los sistemas solares térmicos mediante modelos neurales (in Spanish), University of Valladolid (2016). https://doi.org/10.35376/10324/21048.

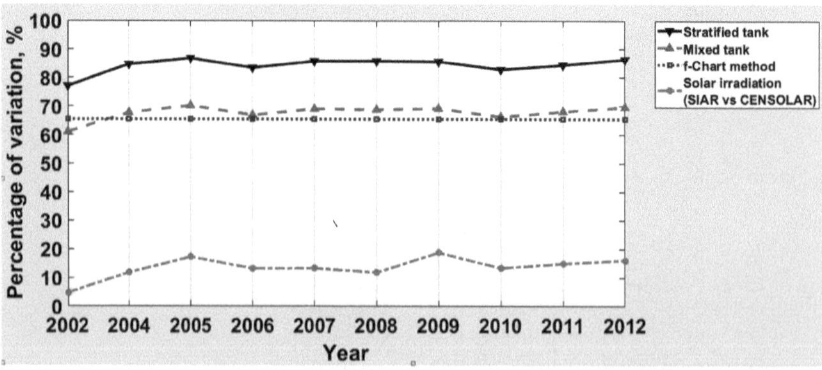

Figure 4.27 Annual performance (10 years) of the solar thermal system. Neural simulation (stratified and mixed tank), f-chart method (mixed tank), and the variation of the solar irradiation with SIAR data vs typical CENSOLAR year. *Modified from F.J. Diez, Optimización del rendimiento en los sistemas solares térmicos mediante modelos neurales (in Spanish), University of Valladolid (2016). https://doi.org/10.35376/10324/21048.*

stratification and with the tank mixed completely at a homogeneous temperature, in addition to the f-chart method using the typical CENSOLAR year. Also, in Fig. 4.27, it the curve of the percentage of the variation of the horizontal global solar irradiation of the database of the SIAR system respect to the CENSOLAR typical year is included.

4.4 Discussion

STSs provide hot water under different solar radiation conditions in every season of the year. However, to meet the required demand under every circumstance, a peripheral power unit is needed [20].

Initially, the validation of the STS working simulation model developed in this study was done through a comparison of the results obtained in its execution and the measured data in the experimental plant during two daily tests (i.e., 10/10/2011 and 11/28/2011, with a working fluid flow of 1 and 1.6 L/min, respectively). The results showed a good adjustment in the prediction model based on the small errors obtained in the stratification temperatures simulated for the eight layers in the tank, with average statistical RMSE values of 0.77°C for the STS dynamic working mode and 0.13°C for the static mode of operation on the first day of the validation (i.e., 10/10/2011) and 0.69°C for the dynamic mode and 0.41°C for the static mode, on the second day of the validation (i.e., 11/28/2011).

As for the results of the simulation of the STS behavior over a year, anomalous or non-coherent results with the daily solar radiation level were checked daily in 2012. None of the results showed that the proposed neural simulation could not be used to assess the operation of the STS over a long period of time and under any circumstances (i.e., weather related or STS operation). Thus, the STS neural simulation model experimented with can be considered valid.

Furthermore, analyzing the STS neural modeling results (Fig. 4.27) as a comparison with the f-chart method application results for the same conditions [i.e., mixed tank, no stratification and working fluid flow of 0.015 kg/(s m^2)], an average deviation of 1.85% was obtained for the solar fraction calculated during a period of 10 years, which also confirms the validity of the STS neural simulation applied over a long period of time. It is also observed that the tank stratification improves the solar fraction by 16.58% over that obtained using the completely mixed tank with a homogeneous temperature.

The f-chart method should be applied with the characteristics of the model of the solar collector tested and the specific flow for the intended method, with which the STS trials should be carried out as the results vary when different flow solar collector models are applied. Thus, for the STS studied in this work, the f-chart method was applied using the Hottel–Whillier–Bliss solar collector model, with working fluid flows of 0.020, 0.015, and 0.009 kg/(s m^2), resulting in solar fractions of 78.25%, 65.48%, and 44.23%, respectively [44]. ANN models formulated in this study consider the different working fluid flows generated when three speeds of the STS circulating bomb were considered. ANN models also considered its influence in the behavior of the heat exchanger and the tank, and in the global performance of the system.

On the other hand, the pipes for the STS primary circuit should be as short as possible to reduce the energy losses [19]. In this study, as that condition was met and there was good thermal insulation in the pipes, the abovementioned thermal losses were not modeled. Thus, the working fluid temperature was considered the same in the pipe at the solar collector outlet (i.e., from collector) and in the inlet of the heat exchanger. Also, the temperature in the outlet of the heat exchanger (i.e., to collector) and the inlet of the solar collector was considered the same.

Bava and Furbo [22] highlighted various points that are considered relevant for the correct modeling of the solar systems:

1. The accuracy of the input data is key for the precise modeling. Thus, in this work the sensors for temperature measurement were calibrated with a different data logger than the one used in the experimental work, and this was considered as the measure pilot.
2. The thermal capacity of the pipes should be considered only for long pipes, because in short pipes its influence is almost negligible. To overcome this simplification, it can be proposed to include in the complete STS model in Simulink MATLAB, a new ANN model for each section of long pipe, which considers the input/output temperatures, flow, and, in the case of exterior circulation, also the ambient temperature should be considered.
3. The coefficient of heat transfer of a heat exchanger should be determined through the measurements instead of using technical specifications. Thus, aspects such as potential contamination due to the quality of the water used and the flow dependency, can be included. This has been the procedure followed in this current work, where all the ANN models have been formulated from performed measurements in the experimental STS, without taking into consideration the technical specifications of its components.

The use of a specific software to achieve an accurate STS model (e.g., TRNSYS) is usually expensive, and requires skilled staff to use it [16]. Furthermore, normally, this kind of tool only allows modeling of systems constituted by components included in the library. In contrast, neural simulation done in this work overcomes these three disadvantages, as it uses a platform for general use, such as Simulink MATLAB, and the formulation of ANN models is done from the performed measurements over the same STS studied components.

Yaïci and Entchev [31] used just one ANN to predict the STS yield, which simulates the stratification temperatures of the preheating tank with six levels with a $\pm 3\%$ error, while the prediction of the heat transfer from the solar collector to the heat exchanger, from the heat transfer to the external tank, and the solar fraction had a $\pm 10\%$ error. Correa-Jullian et al. [32] achieved better results for the simulation through RNN and LSTM (i.e., the lowest RMSE scores 1.27°C for temperature sequence predictions and relative prediction errors of 3.45% for single value predictions), than those obtained with ANNs. In this work, the design of the STS model with Simulink MATLAB, using various ANN models for the tank for each working modes (three ANNs for the dynamic mode and one ANN for the static mode), improves significantly the results of the simulations of stratification temperatures of the tank, with RMSE values of 1.37°C and 0.05°C, being the highest and lowest errors obtained, respectively.

4.5 Conclusions

ANN modeling through experimental STS and subsequent simulation allows its performance to be achieved (which is assessed with the solar fraction variable) and temperatures to be developed (working fluid and water inside the storage tank at different levels), with variable climatic conditions (which depend from solar irradiation and ambient temperature), considering different designs (type of installation) and operation mode (dynamic or static modes).

From the literature it can be seen that there is a scientific trend to apply ANNs to model the components that constitute the STS, using for the training measured data during the operation, organized with time series. In this way, Ghritlare and Prasad [50] reviewed ANN applications to predict the performance of the solar collector and identify some of the gaps in knowledge that remain.

In this work, it ANN models built for the main STS components that work simultaneously has been achieved, with the objective of simulating the complete STS operation, over the short and long term. Thus, a flexible model was developed that can be applied to any STS, and which allows to evaluate the performance through variations in the temperature inside the tank.

Ambient variables, solar irradiance and ambient temperature, and consumption variables, water outlet flow, and cold water inlet temperature in the tank are the input data for the STS neural simulation model. In the case of the solar irradiance, meteorological data are processed to determine its value over an inclined solar collector.

New ANN models, formulated for the STS components, model its operation in the short term (e.g., 1 minute) with measured data for the ambient and consumption variables. Thus, an ANN model has been developed for the flat-plate solar collector, another ANN model for the internal heat exchanger, and four ANN models for the stratified tank, working with dynamic mode (three ANN models for the following cases charge + discharge, charge + no discharge, and no charge + discharge) and with static mode (one ANN model for the case of no charge + no discharge), which increase significantly the precision of the simulation. Also, there is no need to use the technical characteristics of the components for it, which could vary (after the installation or aging).

The most precise way to estimate the STS performance is with a detailed simulation with computer, as was done in this work. Furthermore, the methodology used in this work covers all the advantages detailed in Ref. [51], such as the comprehensive system organization, deep comprehension of the operation, components optimization, estimation of the amount of energy, estimation of temperature variations, and estimation of the effects that will have a change of design variables. In addition, prototype building is not required as the neural simulation has the option of training ANNs which represent the STS components, through theoretical models instead of using measured data in the experimental installations during the operation.

In contrast, neural simulation overcomes the limitation that general simulations have (they only take into account the thermal system behavior) as it uses measured data of the STS components operation. Indeed, ANN models also take into account mechanical and hydraulic considerations in real situations of STS operation (e.g., installation faults, obstructed pipes or with leaks, incrustations, insufficient thermal insulation, control system failures, and different working fluid flows), which are included in the measured and used data for the training of ANN models for the STS components.

STSs are a mature technology, with many different models of collectors and system settings that can be used in different climates [3], thus resulting in a methodology that adapts to different configurations that also could be extended to other renewable energy systems.

Acknowledgments

L.M. Navas-Gracia would like to thank the Ministry of Science, Innovation and Universities, Spain, for the "Estancias de Movilidad de Profesores e Investigadores en Centros Extranjeros de Enseñanza Superior e Investigación" grant program. L.M. Navas-Gracia wishes also to acknowledge Harper Adams University (United Kingdom) and Dr. Laura Vickers for supporting the development of this work.

References

[1] G. Faninger, Solar hot water heating systems, in: A. Sayigh (Ed.), Comprehensive Renewable Energy, Vol. 3, Elsevier Ltd., 2012, pp. 419–447. Available from: https://doi.org/10.1016/B978-0-08-087872-0.00312-7.

[2] M.R. Islam, K. Sumathy, S.U. Khan, Solar water heating systems and their market trends, Renewable and Sustainable Energy Reviews 17 (2013) 1–25. Available from: https://doi.org/10.1016/j.rser.2012.09.011.

[3] K. Hudon, Solar energy – water heating, in: T.M. Letcher (Ed.), Future energy. Improved, Sustainable and Clean Options for our Planet, 2nd (ed.), Elsevier Science, 2014, pp. 433–451. Available from: https://doi.org/10.1016/B978-0-08-099424-6.00020-X.

[4] A. Jamar, Z.A.A. Majid, W.H. Azmi, M. Norhafana, A.A. Razak, A review of water heating system for solar energy applications, International Communications in Heat and Mass Transfer 76 (2016) 178–187. Available from: https://doi.org/10.1016/j.icheatmasstransfer.2016.05.028.

[5] R. Shukla, K. Sumathy, P. Erickson, J. Gong, Recent advances in the solar water heating systems: a review, Renewable and Sustainable Energy Reviews 19 (2013) 173–190. Available from: https://doi.org/10.1016/j.rser.2012.10.048.

[6] S. Sadhishkumar, T. Balusamy, Performance improvement in solar water heating systems – a review, Renewable and Sustainable Energy Reviews 37 (2014) 191–198. Available from: https://doi.org/10.1016/j.rser.2014.04.072.

[7] Z. Wang, W. Yang, F. Qiu, X. Zhang, X. Zhao, Solar water heating: from theory, application, marketing and research, Renewable and Sustainable Energy Reviews 41 (2015) 68–84. Available from: https://doi.org/10.1016/j.rser.2014.08.026.

[8] S. Jaisankar, J. Ananth, S. Thulasi, S.T. Jayasuthakar, K.N. Sheeba, A comprehensive review on solar water heaters, Renewable and Sustainable Energy Reviews 15 (6) (2011) 3045–3050. Available from: https://doi.org/10.1016/j.rser.2011.03.009.

[9] M. Thirugnanasambandam, S. Iniyan, R. Goic, A review of solar thermal technologies, Renewable and Sustainable Energy Reviews 14 (1) (2010) 312–322. Available from: https://doi.org/10.1016/j.rser.2009.07.014.

[10] C.E.C. Nogueira, M.L. Vidotto, F. Toniazzo, G. Debastiani, Software for designing solar water heating systems, Renewable and Sustainable Energy Reviews 58 (2016) 361–375. Available from: https://doi.org/10.1016/j.rser.2015.12.346.

[11] S.A. Klein, W. A. Beckman, F-Chart Software. <http://fchartsoftware.com/fchart>, 2021.

[12] Valentin Software, T*SOL.<http://www.valentin-software.com/en/products/solar-thermal/14/tsol>, 2021.

[13] Vela Solaris, Polysun. <https://www.velasolaris.com/>, 2021.

[14] NRCan Natural Resources Canada, RETScreen Expert. <https://www.nrcan.gc.ca/energy/retscreen/7465>, 2021.

[15] TRNSYS, Transient System Simulation Tool. <http://www.trnsys.com>, 2021.

[16] R.L. Shrivastava, V. Kumar, S.P. Untawale, Modeling and simulation of solar water heater: a TRNSYS perspective, Renewable and Sustainable Energy Reviews 67 (2017) 126–143. Available from: https://doi.org/10.1016/j.rser.2016.09.005.

[17] M.J.R. Abdunnabia, K.M.A. Alakder, N.A. Alkishriwi, S.M. Abughres, Experimental validation of forced circulation of solar water heating systems in TRNSYS, Energy Procedia 57 (2014) 2477–2486. Available from: https://doi.org/10.1016/j.egypro.2014.10.257.

[18] L.M. Ayompe, A. Duffy, S.J. McCormack, M. Conlon, Validated TRNSYS model for forced circulation solar water heating systems with flat plate and heat pipe evacuated tube collectors, Applied Thermal Engineering 31 (8–9) (2011) 1536–1542. Available from: https://doi.org/10.1016/j.applthermaleng.2011.01.046.

[19] L.M. Ayompe, A. Duffy, Analysis of the thermal performance of a solar water heating system with flat plate collectors in a temperate climate, Applied Thermal Engineering 58 (1–2) (2013) 447–454. Available from: https://doi.org/10.1016/j.applthermaleng.2013.04.062.

[20] N. Khordehgah, V. Guichet, S.P. Lester, H. Jouhara, Computational study and experimental validation of a solar photovoltaics and thermal technology, Renewable Energy 143 (2019) 1348–1356. Available from: https://doi.org/10.1016/j.renene.2019.05.108.

[21] C.N. Antoniadis, G. Martinopoulos, Optimization of a building integrated solar thermal system with seasonal storage using TRNSYS, Renewable Energy 137 (2019) 56–66. Available from: https://doi.org/10.1016/j.renene.2018.03.074.

[22] F. Bava, S. Furbo, Development and validation of a detailed TRNSYS-Matlab model for large solar collector fields for district heating applications, Energy 135 (2017) 698–708. Available from: https://doi.org/10.1016/j.energy.2017.06.146.

[23] C. Artur, D. Neves, B.C. Cuamba, A.J. Leão, Comparison of two dynamic approaches to modelling solar thermal systems for domestic hot water, Sustainable Energy Technologies and Assessments 30 (2018) 292–303. Available from: https://doi.org/10.1016/j.seta.2018.10.012.

[24] Chr Lamnatou, J.D. Mondol, D. Chemisana, C. Maurer, Modelling and simulation of Building-Integrated solar thermal systems: behaviour of the coupled building/system configuration, Renewable and Sustainable Energy Reviews 48 (2015) 178–191. Available from: https://doi.org/10.1016/j.rser.2015.03.075.

[25] Chr Lamnatou, J.D. Mondol, D. Chemisana, C. Maurer, Modelling and simulation of Building-Integrated solar thermal systems: behaviour of the system, Renewable and Sustainable Energy Reviews 45 (2015) 36–51. Available from: https://doi.org/10.1016/j.rser.2015.01.024.

[26] K. Zelzouli, A. Guizani, C. Kerkeni, Numerical and experimental investigation of thermosyphon solar water heater, Energy Conversion and Management 78 (2014) 913–922. Available from: https://doi.org/10.1016/j.enconman.2013.08.064.

[27] A.A. Hachicha, B.A.A. Yousef, Z. Said, I. Rodríguez, A review study on the modeling of high-temperature solar thermal collector systems, Renewable and Sustainable Energy Reviews 112 (2019) 280–298. Available from: https://doi.org/10.1016/j.rser.2019.05.056.

[28] Ye Yerdesh, Z. Abdulina, A. Aliuly, Ye Belyayev, M. Mohanraj, A. Kaltayev, Numerical simulation on solar collector and cascade heat pump combi water heating systems in Kazakhstan climates, Renewable Energy 145 (2020) 1222–1234. Available from: https://doi.org/10.1016/j.renene.2019.06.102.

[29] F. Mosallat, T. Elmekkawy, D.L. Friesen, T. Molinski, S. Loney, E.L. Bibeau, Modeling, simulation and control of flat panel solar collectors with thermal storage for heating and cooling applications, Procedia Computer Science 19 (2013) 686–693. Available from: https://doi.org/10.1016/j.procs.2013.06.091.

[30] M. Dongellini, S. Falcioni, G.L. Morini, Dynamic simulation of solar thermal collectors for domestic hot water production, Energy Procedia 82 (2015) 630–636. Available from: https://doi.org/10.1016/j.egypro.2015.12.012.

[31] W. Yaïci, E. Entchev, Performance prediction of a solar thermal energy system using artificial neural networks, Applied Thermal Engineering 73 (1) (2014) 1348–1359. Available from: https://doi.org/10.1016/j.applthermaleng.2014.07.040.

[32] C. Correa-Jullian, J.M. Cardemil, E. López Droguett, M. Behzad, Assessment of Deep Learning techniques for prognosis of solar thermal systems, Renewable Energy 145 (2020) 2178–2191. Available from: https://doi.org/10.1016/j.renene.2019.07.100.

[33] A.H. Elsheikh, S.W. Sharshir, M.A. Elaziz, A.E. Kabeel, W. Guilan, Z. Haiou, Modeling of solar energy systems using artificial neural network: a comprehensive

review, Solar Energy 180 (2019) 622–639. Available from: https://doi.org/10.1016/j.solener.2019.01.037.

[34] InfoRiego, ITACyL Agricultural Technological Institute in Castile and León, Spain, Meteorological information (in Spanish). <http://www.inforiego.org>, 2021.

[35] M. Iqbal, An Introduction to Solar Radiation, Academic Press, 1983. Available from: https://doi.org/10.1016/B978-0-12-373750-2.X5001-0.

[36] R.G. Allen, L.S. Pereira, D. Raes, M. Smith, Crop evapotranspiration: guide-lines for computing crop water requirements. In FAO irrigation and drainage paper, No. 56. Food and Agriculture Organization of the United Nations, 1998. http://www.fao.org/3/X0490E/x0490e00.htm.

[37] J.A. Duffie, W.A. Beckman, Solar Engineering of Thermal Processes, 4th ed., John Wiley & Sons., 2013. Available from: https://doi.org/10.1002/9781118671603.

[38] S.A. Kalogirou, Solar energy engineering: processes and systems, 2nd ed., Academic Press., 2014. Available from: https://doi.org/10.1016/C2011-0-07038-2.

[39] B.Y.H. Liu, R.C. Jordan, Daily insolation on surfaces tilted towards the equator, ASHRAE Journal 3 (1961) 53–59.

[40] A. De Miguel, J. Bilbao, R. Aguiar, H. Kambezidis, E. Negro, Diffuse solar irradiation model evaluation in the North Mediterranean Belt area, Solar Energy 70 (2) (2001) 143–153. Available from: https://doi.org/10.1016/S0038-092X(00)00135-3.

[41] F.J. Diez, Optimización del rendimiento en los sistemas solares térmicos mediante modelos neurales (in Spanish), University of Valladolid. (2016). Available from: https://doi.org/10.35376/10324/21048.

[42] H.B. Demuth, M.H. Beale, M.T. Hagan, Neural network toolbox: user's guide MATLAB, In The MathWorks (2017).

[43] F.J. Diez, L.M. Navas-Gracia, L. Chico-Santamarta, A. Correa-Guimaraes, A. Martínez-Rodríguez, Prediction of horizontal daily global solar irradiation using artificial neural networks (ANNs) in the Castile and León region, Spain, Agronomy 10 (1) (2020) 96. Available from: https://doi.org/10.3390/agronomy10010096.

[44] F.J. Diez, L.M. Navas-Gracia, A. Martínez-Rodríguez, A. Correa-Guimaraes, L. Chico-Santamarta, Modelling of a flat-plate solar collector using artificial neural networks for different working fluid (water) flow rates, Solar Energy 188 (2019) 1320–1331. Available from: https://doi.org/10.1016/j.solener.2019.07.022.

[45] F.J. Diez, L.M. Navas, A. Martínez,). Prediction and identification of the evolution of solar-heated hot water temperature in a stratified tank in static mode using artificial neural networks, Applied Solar Energy (English Translation of Geliotekhnika) 52 (3) (2016) 183–192. Available from: https://doi.org/10.3103/S0003701X16030051.

[46] DGS German Solar Energy Society, Planning and Installing Solar Thermal Systems: A Guide for Installers, Architects and Engineers, 2nd ed., Earthscan, 2010. Available from: https://www.routledge.com/Planning-and-Installing-Solar-Thermal-Systems-A-Guide-for-Installers-Architects/DGS/p/book/9781844077601.

[47] W.A. Beckman, S.A. Klein, J.A. Duffie, Solar Heating Design by the F-Chart Method, John Wiley & Sons, 1977.

[48] J.J. Mutch, Residential water heating, fuel consumption, economics and public policy (Rand, Ed.), Report R (1974) 1498.

[49] CENSOLAR Center for Solar Energy Studies, Instalaciones de energía solar térmica. Pliego de condiciones técnicas de instalaciones de baja temperatura (in Spanish) (PET-REV), IDAE Institute for Diversification and Saving of Energy, 2009, pp. 100. https://www.idae.es/uploads/documentos/documentos_5654_ST_Pliego_de_Condiciones_Tecnicas_Baja_Temperatura_09_082ee24a.pdf.

[50] H.K. Ghritlahre, R.K. Prasad, Application of ANN technique to predict the performance of solar collector systems – a review, Renewable and Sustainable Energy Reviews 84 (2018) 75–88. Available from: https://doi.org/10.1016/j.rser.2018.01.001.

[51] A. Athienitis, S.A. Kalogirou, L. Candanedo, Modeling and simulation of passive and active solar thermal systems, in: A. Sayigh (Ed.), Comprehensive Renewable Energy, Vol. 3,). Elsevier Ltd., 2012, pp. 357–417. Available from: https://doi.org/10.1016/B978-0-08-087872-0.00311-5.

CHAPTER FIVE

Solar energy modelling and forecasting using artificial neural networks: a review, a case study, and applications

Youness El Mghouchi
Moulay Ismail University, National School of Arts and Crafts (ENSAM), Meknes, Morocco

Contents

5.1 Introduction

Over the past and during the current century, it has become clear that temperature data show a general warming trend and it should be noted that most scientist experts on the subject of climate change believe

Artificial Neural Networks for Renewable Energy Systems and Real-World Applications.
DOI: https://doi.org/10.1016/B978-0-12-820793-2.00009-4

that the effects of anthropogenic drivers are huge and are the dominant reason for the detected climate change over the past 15 years [1]. Anthropogenic drivers, which include human transport, industrial, and agricultural activities, are causing greenhouse gas emissions related to the poor management of energy use. This could be because of several problems: supply and demand imbalance, the unsustainability of conventional energy resources, and the drastic consequences of global warming, which could adversely affect directly human health and/or humankind's existence [2].

Accordingly, renewable energies have been recommended in the last few decades as suitable alternative sources of energy supply, thanks to being clear, free, environmentally friendly, and broadly accessible energy forms across the globe. Therefore, there has been growing awareness and a great deal of attention from the world's states to renewable energy resources. In this regard, countries are now giving tremendous efforts to assign a high priority to different renewable energy resources by investing intensively in this field of cleaner energy production and also decreasing the use of fossil fuels.

The term 'renewable energy' includes solar energy, wind energy, biomass, hydroelectric power, and wave power. Solar energy, which reaches the Earth in the form of solar radiation after crossing several million kilometers, is power that could be harnessed through the sun rays falling on solar-powered devices. The solar radiation intensity that arrives at the Earth is equal to 10,000 times the annual total energy consumption in the world. On average, the solar intensity received is about 1700 $kWh/m^2/year$ [3].

Solar energy has gained increasing scientific attention during recent years. The value information of this energy is very important for researchers, engineers, designers, and architects. It is also a fundamental parameter for efficiently determining irrigation water needs and potential crop yields, and for realistic studies of the solar effects on the Earth's climate system. However, in many areas, especially in underdeveloped and developing countries, measures of solar radiation are limited or not available. To extend these limited data, several techniques and methods for forecasting and modeling of solar radiation with certain levels of uncertainty are of paramount importance [4].

The input variable selection (IVS) scientific area, which is a field of great scientific interest, is considered for modeling of solar energy and is crucial in generating accurate estimates. This modeling, which serves to compute the components of the solar radiation at a given position of

Earth, is given under both empirical and parametrical forms. In literature the authors were only able to find solar-based model functions of astronomical parameters and/or functions of climatic conditions and meteorological variables. These latter categories can be classified into sunshine duration-based models [5,6], temperature based-models [7], cloud factor-based models [8], and models based on multiple meteorological parameters [9,10].

In general, semiparametrical models [11,12] have been extensively used in this area, by correlating the total solar radiation with other easily measurable parameters such as meteorological data, astronomical parameters, and geographical coordinates, and by applying compendious mathematical formulas. Thanks to their simple expressions, they are operational and efficient for several applications. However, the accuracy of such models is quite debatable in some cases because it primarily depends on the solar cycle during which the model was developed [13] and on the site's location where the model was validated.

Besides, to offer more accuracy, reliability, and convenience in the estimation of solar radiation, the application of artificial and computational intelligence techniques has attracted great attention over the last two decades, including artificial neural networks (ANNs) [14–16], support vector machines (SVMs) [17,18], genetic programming [19], autoregressive moving average [20], satellite-based models [21], etc.

During these past 15 years, ANNs have become as effective correlative techniques in several fields and their applications have become of interest, especially for predicting solar radiation. A detailed review of solar radiation forecasting using ANN techniques is presented in detail in Ref. [22].

Furthermore, in certain cases, various hybrid approaches of ANNs, and their derived techniques, have been established. Some researchers have considered IVS techniques [or feature selection (FS)] combined with prediction techniques like ANN, SVM, and/or decision tree [23,24].

In literature, several IVS techniques are described and evaluated, including, World Competitive Contests (WCC) [25], League Championship Algorithm (LCA) [26], Genetic Algorithm (GA) [27], Particle Swarm Optimization (PCO) [28], Ant Colony Optimization (ACO) [29], Imperialist Competitive Algorithm (ICA) [30], Learning Automata (LA) [31], Heat Transfer Optimization (HTO) [32], Forest Optimization Algorithm (FOA) [33], Discrete Symbiotic Organisms Search (DSOS) [34], Cuckoo Optimization Algorithm (COA) [35], etc. Besides, for each IVS technique, only one set of inputs can be obtained as an optimal combination for predicting the considered output parameter.

However, to the best of our knowledge, there have been no systematic studies in real-world solar forecast searching for all possible combinations of inputs and then applying them to the considered prediction technique. Multiple authors have obtained different sets of input parameters and considered them in the prediction phase for drawing their conclusions.

In this chapter, an intelligent approach for searching for the best combinations of inputs and applying them in solar radiation forecasting is presented and evaluated as a case study. Specifically, this approach focuses on three perspectives that were not considered in previous studies of IVS. First, it investigates the performance of hybrid IVS-ANNs by combining IVS and ANNs for daily and monthly solar radiation forecasting using local weather information from 35 sites with different climate characteristics. Second, it selects the best combinations of inputs that could give the considered output parameter with the best possible correlations, resemblance, and approximations. Third, it investigates to predict, with the highest possible accuracy, all other related meteorological parameters like temperature, humidity, and wind speed.

In addition, as a review analysis, different timescales of solar radiation intensities (instantaneous, hourly, daily, and monthly) based on multiple geographical, astronomical, and climatological input parameters are presented. A review of solar energy applications for both human health and engineering is briefly summarized. These cleaner production applications include, but are not limited to, producing vitamin D, solar fuel, solar water heaters, solar air-conditioning, solar desalination, photovoltaic (PV) systems, solar power water pumping, hybrid PV/thermal, and solar concentrating power generation.

In the remainder of this chapter, Section 5.2 describes the solar radiation modeling and the methodology followed. In Section 5.3, we present the local weather information used, and statistical analysis considered. Section 5.4 shows the study results, and the discussion. Section 5.5 gives an overview of the applications of solar energy and its conversion systems. Conclusions are drawn in Section 5.6.

5.2 Solar radiation modeling

Many researchers have tested existing models or developed new ones to predict the all-time series of solar intensity falling on an area of the Earth for their city/location and/or their state. A frequently most debated issue about the proposed correlation models in the forecasting of solar intensity is the selection of the number and category of meteorological, geographical, and astronomical parameters. In Table 5.1 we introduce the most frequently adopted parameters by researchers worldwide. Despite geographical, astronomical, and meteorological data being the most commonly considered by several authors, no standard is considered because the best set of inputs and type of parameters is location dependent.

5.2.1 Solar constant and extraterrestrial radiation

The solar constant, as defined, is the total quantity of solar energy per unit of time at the averaged Earth—sun distance, reached by a unit of area of a surface perpendicular to the sun's rays (the normal sun direction) outside the atmosphere. It can be calculated from the total power emitted by the surface of the sun, and is about 63.11 MW/m^2 [44]. By relating this radiant emittance to the square meter rate of the Earth's radius r_s (about $6.9598.10^8$ m) to the sun—Earth distance r_0 (about $1.496.10^{11}$ m), we find:

$$I_{sc} = M\left(\frac{r_s}{r_0}\right)^2 = 63.11.10^6\left(\frac{6.9598.10^8}{1.4959789.10^{11}}\right)^2 \approx 1367 \quad \frac{W}{m^2} \tag{5.1}$$

Table 5.1 Frequently adopted parameters in the forecasting of solar energy.

Parameter	References
Air temperature	[36]
Relative humidity	
Sunshine duration	
TAO (Top Out Atmosphere)	[37]
Solar angles	
Clearness index	[38]
Altitude	[39]
Atmospheric pressure	
Day length	
Time period	
Atmospheric transparency	
Albedo at ground level	
Turbidity factor	[40]
Rainfall	[41]
Climatic conditions	[42]
Wind speed	[43]

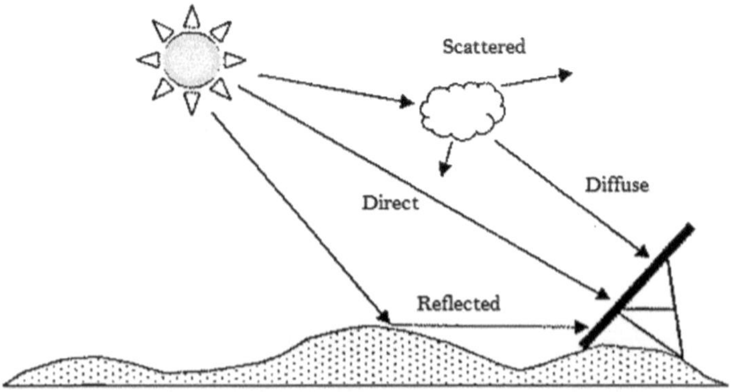

Figure 5.1 Global solar radiation components: direct, diffuse, and reflected radiations.

In addition, as the Earth's orbit is elliptical, this intensity varies in relation with the sun–Earth distance variations throughout the year. Consequently, the solar constant can be corrected to:

$$I_0 = E_0.I_{sc} \tag{5.2}$$

where E_0 is the correction of the Earth–sun distance.

5.2.2 Instantaneous and hourly solar radiation models

In this category, solar radiation is linked directly and explicitly to geographical coordinates and solar constant.

The solar radiation arrives on the Earth's surface with three components: the direct flux I (W/m^2), the diffuse flux D (W/m^2), and the reflected flux R (W/m^2). All these fluxes form the global solar radiation G (W/m^2). The direct flux is the part of the radiation that directly arrives on the Earth's surface and is also called the beam radiation. It does not depend on the part of radiation absorbed and reflected by the atmosphere components (aerosols, water vapor, ozone, etc.). The diffuse flux is the portion of scattered radiation in the atmosphere. It does not have a particular direction. The reflected flux is considered null for horizontal surfaces (see Fig. 5.1).

Descriptions and details of the instantaneous and hourly solar radiation fluxes can be found in Refs. [45,46].

5.2.3 Modeling of daily global solar radiation (DGSR) and monthly average global solar radiation (MAGSR) based on ANN techniques

Knowing the insolation received during a given day and/or a given month (in average) is crucial for assessing the effectiveness and efficiency of several cleaner energy devices, like solar energy systems and other related engineering applications. In the literature, several models have been recently elaborated for forecasting, with high possible accuracy, both the daily and monthly average global solar radiation intensities (DGSR and MAGSR, respectively) based on multiple intelligent techniques known as machine learning methods [47,48]. Developed models were given as functions of multiple geographical, astronomical, and climatological combinations. The clearness index is one of these considered parameters. This parameter is defined as the ratio of the measured DGSR (denoted also as H) to daily extraterrestrial solar radiation H0 (or TOA). It indicates the density of clouds in the sky and its impact on the incoming solar radiation. Therefore, it gives both the level of availability of the solar radiation and changes in the climatic conditions in a considered environment [49].

Here in this chapter, the considered approach in predicting both the DGSR and MAGSR was elaborated based on ANN techniques [50]. The nine different inputs that were considered are presented in Table 5.2.

For these considered inputs, **511** possible combinations should be compared for determining the optimal combination(s) that could provide the best accuracy in predicting the DGSR or MAGSR.

The number of possible input combinations is determined using the following equation:

$$\mathrm{Comb} = \sum_{p=1}^{n} C_n^p = \sum_{p=1}^{n} \frac{n!}{(n-p)!p!} \tag{5.3}$$

where n represents the number of considered inputs.

Table 5.2 The considered inputs and output parameters.

Input number	**Parameter**	**Unit**
Input 1	KI	–
Input 2	$T_{average}$	°C
Input 3	T_{max}	°C
Input 4	T_{min}	°C
Input 5	ΔT	°C
Input 6	TR	-
Input 7	RH	%
Input 8	Ws	m/s
Input 9	TOA	kWh/m^2
Outputs	DGSR and MAGSR	kWh/m^2

Additionally, **511** results were found and stored for the nine inputs considered. These results should be compared for determining the best combinations of inputs that could provide DGSR and/or MAGSR with maximum possible correlations and approximations. For each combination of inputs, the network is created and launched, then the corresponding data are imported, and the target, training, and testing data are selected. An algorithm under MATLAB® software was elaborated and used to statistically assess the obtained predicted values by comparing them with the measured values, then the result is displayed and shows the best combinations of inputs in predicting both the DGSR and the MAGSR. The total times for computing and getting the result for both the DGSR and the MAGSR were about 36 hours and 3 hours, respectively.

The procedure considered is well explained in the flowchart in Fig. 5.2. This later was elaborated using UML activity diagrams (LUCIDCHART). This flowchart demonstrates the great adaptations applied to the used traditional ANN technique.

As shown in Fig. 5.2, the main steps in this approach are illustrated and described as following:

1. Start the algorithm.
2. Import the inputs and output data.
3. Compute the total number of combinations based on the data size provided and using Eq. (5.3).
4. Start a first loop based on the size of the provided data, K1.

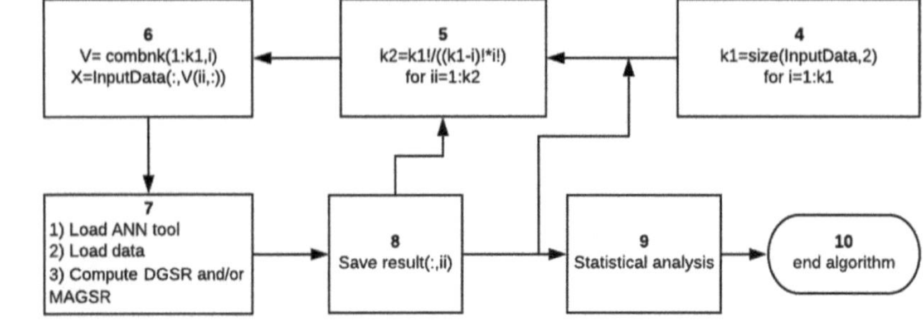

Figure 5.2 Flowchart for the proposed approach.

5. Compute the number of combinations for each ith considered size, and then start a second loop for each value of K2.
6. Use the combnk(V, K) function for producing matrix with K columns.
7. Load ann tool and load data in for computing the considered output parameter.
8. Save the computed values of DGSR (or MAGSR) and go for the next iteration.
9. After getting the predicted values by all considered combinations, the final result is imported by a second algorithm in which the statistical analysis is done.
10. The best combinations of inputs are determined, and then the algorithm is ended.

Moreover, by applying the least square regression (LSR) [51,52] to the found best combinations of inputs, several new linear formulas between the considered output (DGSR or MAGSR) and the obtained best combinations of inputs were developed and summarized in such corresponding tables.

$$\text{Output} = a_0 + a_1\text{KI} + a_2\text{Ta} + a_3\text{Tmax} + a_4\text{Tmin} + a_5\Delta\text{T} + a_6\text{TR} + a_7\text{Rh} + a_8\text{Ws} + a_9\text{TOA} \quad (5.4)$$

Linear relationships were considered because of their capability for computing both the considered output and other related input parameters. For example, if we found the two best combinations, the first with two input parameters (KI and TOA as an example) and the second with three (KI, Ta, and TOA), considering the first relationship, we could predict the output considered parameter, and considering the second relationship, the Ta is then conducted.

5.3 Used data and statistical analysis

5.3.1 Local weather information

Data from 35 Moroccan cities were considered and employed. They were provided in hourly, daily, and monthly averaged time series. The data provided were recorded by the Solar Energy Laboratory of the Faculty of Sciences of Tetouan, the Solar Energy Laboratory of the Faculty of

Sciences of Rabat, and the websites http://www.rp5.ru and http://www.ogimet.com/gsynres.phtml, from 01/02/2005 to 01/01/2018. Table 5.2 illustrates the corresponding inputs and outputs.

5.3.2 Statistical analysis

For assessing and finding the best combination of inputs, a statistical analysis was done. The methodology presented by Badescu was adopted [53]. It is based on a performance score (φ) that is defined as:

$$\varphi = \text{rank(MBE)} + \text{rank(RMSE)} + \text{rank(TS)} + \text{rank}(R^2) + \text{rank(WIA)} + \text{rank(SBF)} \tag{5.5}$$

Higher values of φ indicate poor accuracy.

The corresponding formulas giving these used indicators are cited below:

$$\text{MBE} = \frac{1}{K}\sum\left(H_p^i - H_m^i\right) \tag{5.6}$$

$$\text{RMSE} = \left(\frac{1}{K}\sum\left(H_p^i - H_m^i\right)^2\right)^{1/2} \tag{5.7}$$

$$\text{TS} = \left[\frac{(K-1)\text{MBE}^2}{\left(\text{RMSE}^2 - \text{MBE}^2\right)}\right]^{1/2} \tag{5.8}$$

$$R^2 = 1 - \frac{\sum\left(H_p^i - H_m^i\right)^2}{\sum\left(H_m^i - \overline{H_m}\right)^2} \tag{5.9}$$

$$\text{SBF} = \frac{\left[\sum\left(H_p^i - \overline{H_p}\right)\left(H_m^i - \overline{H_m}\right)\right]}{\sum\left(H_m^i - \overline{H_m}\right)^2} \tag{5.10}$$

$$\text{WIA} = 1 - \frac{\left[\sum\left(H_p^i - H_m^i\right)^2\right]}{\sum\left[\left|H_p^i - \overline{H_m}\right| + \left|H_m^i - \overline{H_m}\right|\right]^2} \tag{5.11}$$

Moreover, for additional overall performance, other indicators are considered such as standard deviation (σ), which presents the level of differences between predicted and measured values, normal root mean square error (NRMSE), mean absolute percentage error (MAPE), mean percentage error (MPE) and Legate's coefficient of efficiency (LCE).

$$\sigma = \left[\frac{K\left(RMSE^2 - MBE^2\right)}{(K-1)}\right]^{1/2} \tag{5.12}$$

$$NRMSE = \frac{\left(\frac{1}{K}\sum\left(H_p^i - H_m^i\right)^2\right)^{1/2}}{\frac{1}{K}\sum\left(H_m^i\right)} \tag{5.13}$$

$$MAPE = \frac{100}{K}\sum\left|\frac{\left(H_p^i - H_m^i\right)}{H_m^i}\right| \tag{5.14}$$

$$MPE = \frac{100}{K}\sum\frac{\left(H_p^i - H_m^i\right)}{H_m^i} \tag{5.15}$$

$$LCE = 1 - \frac{\sum\left(H_p^i - H_m^i\right)^2}{\sum\left(H_m^i - \overline{H_m}\right)^2} \tag{5.16}$$

In these formulas, K represents the total number of measures, and H_p^i, H_m^i, and $\overline{H}$ are the ith forecasted value, ith measured value, and mean value, respectively.

5.4 Results and discussions

5.4.1 Best combinations of inputs in modeling and forecasting of DGSR

The results presented in this subsection were well assessed and evaluated. In Fig. 5.3, the best combinations of inputs we found are shown. The 128 best combinations in predicting the DGSR were found. The TOA and KI, in these combinations, were repeated 128 times. This indicates that, from all considered input variables, TOA and KI are considered as indispensable inputs to the DGSR modeling. Other inputs appeared in these combinations 64 times, indicating their insignificant impacts on DGSR predictions in the missing of TOA and KI. Furthermore, no combination gave DGSR with only one input parameter.

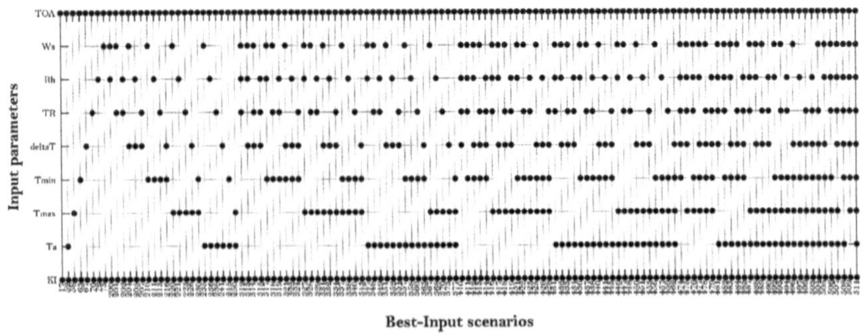

Figure 5.3 Best combinations of inputs recommended for modeling of the DGSR.

As shown in this figure, the first best combination is a function of KI, Ta, ΔT, TR, and TOA. The second is a function of KI, Ta, Tmax, TR, RH, Ws, and TOA. The third is a function of KI, Ta, Tmax, TR, Ws, and TOA. The fourth is a function of KI, Ta, TR, RH, and TOA. However, the first and fourth models predict DGSR with five input variables, the third predicted it with six, while the second gives DGSR with seven input variables.

We believe that the found best combinations of inputs for DGSR modeling are very useful for forecasting other related meteorological inputs such as averaged temperature, Tmax, Tmin, humidity, and wind speed. This could be elaborated using an interface graphic or a software application under Windows and/or phone apks. In this interface, by giving information about at least one input variable like KI, we could predict all other related parameters. We can use KI as a predictor input variable because of its simplicity and accessibility.

The linear formulas between DGSR and the best combinations of inputs are presented in Ref. [50]. Considering these formulas and based on information given by an empirical model that may estimate the solar radiation value using only the Julian day (day number between 1 for January 1st and 365 for December 31st) as input, we could forecast the DGSR and other related parameters like averaged temperature, humidity, and wind speed.

The first best model that was established presents DGSR with three associated inputs: TOA, a relationship or expression involving Ta and its other related indicators (Tmax, Tmin, ΔT, etc.), and KI. The corresponding statistical indicator values are: 0.992 for R^2, −0.203 for MBE, 0.302 for RMSE, 102,553 for TS, 0.926 for SBF, and 0.994 for WIA. The values of R^2, SBF, and WIA are closer to 1, indicating incomparable correlations and linear formulas between the values computed by this model and measured DGSR. Small values of indicators of dispersion indicated a great approximation between the predicted values and the measured ones. In addition, for all these found best combinations, values of MBE (%), RMSE (%), and σ (%) are around, respectively, -0.002 to -0.003, 0.002 to 0.004, and 0.002 to 0.003. As can be seen, values of MBE and RMSE are close to 0, which indicates the best possible likeness between the measured values and those predicted (see Fig. 5.4).

Furthermore, in Morocco, we have elaborated and tested a new model [54,55] for forecasting the instantaneous and then the daily solar radiation for clear sky conditions function of a turbidity factor, which represents dispersion, diffusion, and transmission phenomena caused by atmospheric components. This factor is given only function of the Julian day (day number: 1 for 1 January, and 365 for 31 December). We could use this model for predicting DGSR on clear sky conditions and, using the linear formulas developed in this work, we can predict other corresponding input variables.

In Morocco, based on the results found for daily predictions, we have elaborated two applications, **IntellIrrig** and **ClimForecast** (see Fig. 5.5). These applications were assigned for agricultural fields, particularly for intelligent irrigation, and for fields linked to the climate, respectively. Using **IntellIrrig** for a given site, we can predict the quantity of water that is required for a considered plant. Besides, with **ClimForecast**, we can predict with high precision the most important parameters in the field of metrology including average temperature, humidity, insolation, wind speed, etc.

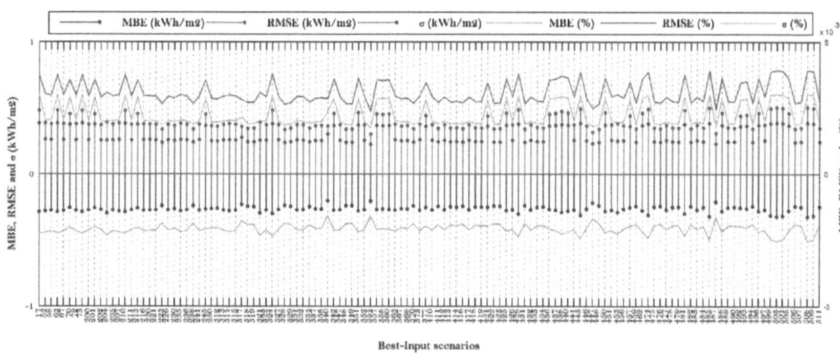

Figure 5.4 Statistical analysis results for the found best combinations of inputs in the forecasting of DGSR.

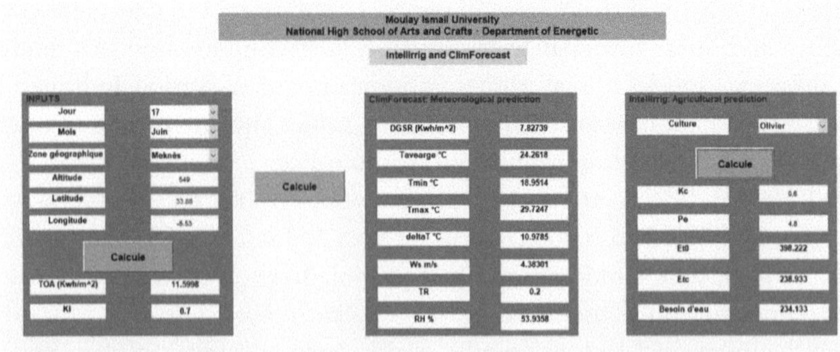

Figure 5.5 The two elaborated applications: IntellIrrig and ClimForecast.

5.4.2 Best combinations of inputs in modeling and forecasting of MAGSR

Several weather stations provide solar radiation data by degrees of availability; where hourly, daily, and monthly values of this parameter are less available, medium available, and strongly available, respectively. Also, by using decomposition techniques [56,57], many researchers have proposed methodes for transforming monthly solar radiation values into daily, hourly, and/or instantaneous values.

In literature, we can also find several research papers that discuss the subject of forecasting MAGSR by integrating certain artificial intelligence techniques. Some researchers have used an adaptive neuro-fuzzy inference (ANFIS) system and other hybrid models covering PSO, GA, the cuckoo search algorithm (CS), and differential evolution (DE) [58,59] as IVS techniques, while other researchers have employed SVM [60].

More specifically, in Morocco, some researchers have considered certain techniques of ANNs like multilayer perceptron (MLP) and radial basis function (RBF) for forecasting MAGSR [61]. The inputs considered in these studies were a mixture of meteorological variables including sunshine duration, Ta, Tmin, Tmax, ΔT, rainfall, TOA, and KI.

The common point between these studies was the utilization of a sole combination of inputs for showing conclusions. Therefore, using the adopted approach and as mentioned previously, the prediction of MAGSR is done by comparing all possible combinations of inputs and by selecting the most relevant one(s) that may give the considered output with maximum possible correlation and very close approximation.

As is shown in Fig. 5.6, the best combinations of inputs in forecasting of MAGSR were found and illustrated. There are 85 best combinations

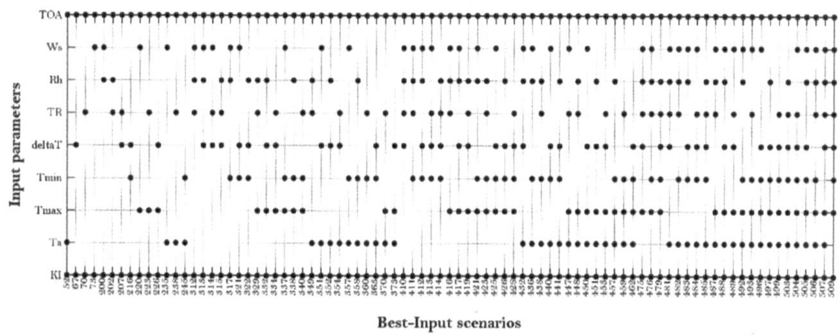

Figure 5.6 Best combinations of inputs for MAGSR modeling.

of inputs may be used in modeling of MAGSR. These combinations are given the function of multiple input variables that appeared 85, 45, 44, 44, 48, 44, 46, 42, and 85 times (Input 1 to Input 9, respectively). This demonstrates also that the TOA and KI are the keys to modeling the MAGSR. Other inputs were repeated 42–48 times, indicating their small impact on MAGSR predictions in the absence of TOA and KI.

Obviously, no singular combination can give the MAGSR with only one input.

To statistically assess the accuracy of the proposed approach in predicting the MAGSR, we have considered the coefficient of determination R^2 (level of confidence is 0.95), which indicates the level of correlations between the predicted and the measured values, and the performance score φ for ordering the results found.

The statistical analysis has been done (see the Appendix, Table 5.A1) and it shows that the first best combination was found for combination **411**, which gives MAGSR function of KI, Tmin, TR, Rh, Ws, and TOA. The corresponding statistical indicator values for this combination are 0.987 for R^2, 3.438 for MBE, 8.087 for RMSE, 9.613 for TS, 0.894 for SBF, and 0.997 for WIA. Moreover, the values of R^2, SBF, and WIA for this model are closer to 1, indicating very ideal correlation, likeness, and the interrelationship between the considered inputs of this combination and MAGSR. Small values of other indicators indicate very good approximations between the values provided by this model and the measured MAGSR. Statistical results for the other best combinations are summarized in the same table. There are some differences in values of the considered statistical indicators, where the values of indicators of dispersion in percentage are very closer to zero, and the values of indicators of performance are very close to 1.

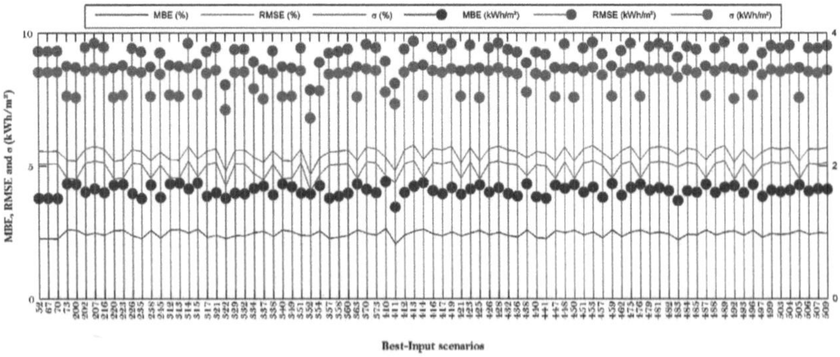

Figure 5.7 Statistical analysis results for the found best combinations of inputs in the forecasting of MAGSR.

In addition, Fig. 5.7 presents the obtained results of MBE, RMSE, and σ in $kWh/m^2/day$ and in percentage. They range, respectively, from 0.8186 to 1.0497, from 1.8686 to 2.3073, and from 1.6162 to 2.0775. Values of these indicators are around 1%–2%, indicating very good approximations between the values predicted and the measured ones. This well indicates that the approach adopted is very useful for good forecasting purposes and the found best combinations may be considered as an excellent way of elaborating parametric models for predicting MAGSR.

Linear formulas for forecasting of MAGSR have been developed and summarized in Table 5.A2 (see the Appendix). There are 85 new linear formulas for forecasting MAGSR function of multiple meteorological and climatological variables. Based on these formulas, we could determine not only MAGSR but also other related variables like monthly averaged values of Ta, ΔT, Rh, and Ws. For example, by considering the four premier established linear formulas which correspond to the combinations 52, 67, 70, and 73, we could simply predict, respectively, values of Ta, ΔT, TR, and Ws. Other variables may be determined by considering other linear formulas and/or by elaborating a graphic interface that may be used by any user without knowledge about solar forecasting.

5.5 Solar energy conversion systems: an overview

A transition is underway around the world toward energy systems powered fully by free, abundant, local, clean, and affordable renewable energy sources from energy systems that are powered by high-priced and unsustainable fossil fuel resources. In the coming years, this transition is expected to improve the quality of life for people, with an absolute reduction in harmful greenhouse gas emissions, and helping move the world forward for both current and future generations.

Applications of solar energy engineering have been widely reviewed for solar space heating and cooling, water heating, cooking, electrical generation, and several other conversion systems of sunlight into a useful energy form. Various applications have been analyzed and discussed for systems powered by solar energy. Laws of thermodynamics, aerodynamics, principles of mechanics of fluids, and physics of heat transfer are considered in describing the thermal and physical properties of solar receiver systems, heat exchangers, and fixed beds. The types of collectors reviewed include flat-plate, tubular, concentrating, fixed, or intermittently turned concentrators, and various technologies of tracking concentrators. Procedures for analyzing solar energy systems, components, and economics have been developed and applied to thermal and electrical storage, energy transport, heat exchangers, and control systems [62,63].

Systems powered by electrical generation could be achieved either directly, using PV technologies, or indirectly, by an accurate method of collecting and optionally concentrating sun rays (CSP) to produce steam that is then utilized for driving turbines and providing electrical power, while the thermal aspects are reviewed for solar heating, water heating, cooling, cooking, and dehumidifying systems.

Solar PV means the direct conversion of sun rays into electrical energy using solar cells. Semiconductor solar cells based on crystalline silicon are the mainstay of the market today, but new technologies of cells based on organic materials, plastics, or thin film are increasingly achieving marketability. Thus, there are monocrystalline modules with up to 20% efficiency and polycrystalline modules with 14% efficiency on the market [64].

Knowledge of solar radiation intensity is the primary consideration in selecting accurate locations and in dimensioning purposes for PV and thermal systems. The electricity output of PV panels and the output of thermal energy systems are directly related to solar radiation and temperature inputs. Local climatic and environmental factors such as ambient temperature, humidity, precipitation, and wind constrain the outputs of these systems. However, these are all secondary effects when considering solar radiation. The total averaged yearly solar radiation intensity changes with latitude at different locations, seasons, time of day, and atmospheric conditions. The amount of power produced by these technologies depends on the quantity and quality of the radiation to which they are exposed.

Therefore, solar radiation, in general, can be converted in three main ways:

- By thermal conversions: solar energy is converted into heat that can be used for the purpose of air-conditioning, for hot water needs, or for electricity production.
- By using the PV effect: solar energy is directly converted into electricity using a variety of solar cell technology like thin-film cells, mono- and polycrystalline cells, organic cells, etc.
- By using the photosynthesis process in the chemical, biological, and agricultural sectors: solar energy is used to generate fuel, hydrogen, glucose, carbohydrates, and oxygen by using water, carbon dioxide, and sunlight. These can be done by several techniques and methods, including: thermolysis of water, thermochemical cycles, cracking of gaseous hydrocarbons, and reforming or gasification of fossil fuel or biomass resources.

For thermal conversions, there is a wide variety of technical means, which are presented within two groups:

Group 1. Stationary systems for direct and diffuse solar radiation and their conversions into thermal energy at low temperatures. These do not require tracking the sun, have simple construction, and require little maintenance. The crucial applications for these devices are in the fields of water heating, building heating, air conditioning, and industrial process heat. This group includes:

- Plane solar collector for water and air heating;
- System of solar plane collector—flow turbine for generating electricity;
- System of accumulated heat solar energy plane collector—flow turbine (solar towers).

Group 2. No stationary systems for solar beam concentrations based on systems of sun tracking. These systems convert the beam radiation into steam at average and high temperatures. They use associations of mirrors and/or lenses for concentrating the direct solar radiation and then producing different forms of useful energy such as heat, electricity, or fuels. In this group, we find:

- Cylindrical parabolic trough collectors;
- Parabolic collectors with different types of engines (steam turbine, Stirling engine, etc.);
- Heliostat fields with a tower;

Table 5.3 Solar energy collectors.

Motion	**Collector type**	**Absorber type**	**Concentration ratio**	**Indicative temperature range (°C)**
Stationary	Flat plate collector (FPC)	Flat	1	30–80
	Evacuated tube collector (etc.)	Flat	1	50–200
	Compound parabolic	Tubular	1–5	60–240
Single-axis	collector (CPC)		5–15	60–300
tracking	Fresnel lens collector (FLC)	Tubular	10–40	60–250
	Parabolic trough collector (PTC)	Tubular	15–45	60–300
	Cylindrical trough collector (CTC)	Tubular	10–50	60–300
Two-axes tracking	Parabolic dish reflector (PDR)	Point	100–1000	100–500
	Heliostat field collector (HFC)	Point	100–1500	150–2000

- Dish concentrating solar power;
- Central solar power systems.

Table 5.3 lists various characteristics of these two groups [65].

For PV conversion systems, there is a variety of technologies used to transform sunlight into electricity. Solar PV cells can be divided, also, into three most promising groups:

Group 1. Crystalline silicon cells: with an efficiency reaching 20%, this technology represents more than 85% of the PV industry productions, through cells and modules based on poly-, mono-, and multicrystalline wafer technology.

Group 2. Thin-film solar cells: with low efficiency, this technology has mainly been proposed for building integrated markets and window-based applications.

Group 3. Concentrated solar PV: in this technology, various curved mirrors and lenses are used to collect and concentrate sun rays onto small, but highly efficient, multijunction solar cells. It requires, generally, sun tracking and cooling systems to increase its efficiency.

For photosynthesis conversion systems, there are different ways used to transform solar energy into glucose, carbon, hydrogen, or, generally, fuel. Solar photosynthesis conversion systems can be divided into the following most-promising categories:

- Thermolysis of water: this technique refers to the thermal decomposition of water into gases (hydrogen and oxygen).
- Thermochemical cycle: this technique is used for the storage and transport of energy (converting solar energy to another storable energy form) based on oxidation/reduction (redox) reactions of a typical metal.
- Cracking of gaseous hydrocarbons: this technique is another method for clean and clear fuel production. It is mainly based on the thermal decarbonization (reduction of carbon content) of natural gas or other hydrocarbons.
- Reforming and gasification of fossil and biomass: In this technique, a thermochemical process is used by converting a carbonaceous feedstock into fuel under a controlled amount of steam.

All these techniques require high temperatures of about 1500K–1700K that can be reached using concentrating solar collectors.

Also, there are many other simple and useful utilizations powered by solar energy, such as:

- Solar night light;
- Solar bag;
- Solar power water pumping system;
- Solar-powered led streetlight with auto intensity control;
- Sun tracking solar panel;
- Solar power charge controller;
- Solar-powered auto irrigation system;
- Solar energy measurement system;
- Solar-powered air conditioning unit;
- Concentrated solar fuel production;
- Solar ventilation air preheating;
- Solar heat for the swimming pool;
- Solar-powered cold-heat using the thermal machine with absorption and/or adsorption functioning;
- Solar food processing based on solar cooking methodologies;
- Solar stills for purifying water;
- Photochemical solar cells for producing hydrogen via the electrolysis of water;

- Solar engines for producing mechanic energy;
- Solar vehicles for providing sustainable transport;
- Solar robotics.

More descriptions and details about different conversion technologies of sunlight can be found in Refs. [66,67].

Furthermore, by considering our field of expertise, particularly solar radiation modeling and forecasting, we believe that sunray exposure may have, in general, an important positive impact on human health. After a careful literature review, we found that there are several benefits to sun rays, which are composed of visible, infrared, and ultraviolet rays, for human health. It could be possible to directly harness energy provided by these rays for producing vitamin D (known as the sunshine vitamin), for increasing people's immunity, for improving heart health, and/or for improving symptoms of mood disorders. Researchers in medical sciences have concluded that regular exposure to sunlight has a huge impact on depression, seasonal affective disorder, and sleep quality, and contributes to the prevention of colon, breast, and prostate cancer, non-Hodgkin lymphoma, multiple sclerosis, hypertension, and diabetes. We believe that a controlled dose of sun rays to patient may destroy undesirable cellular biology. We also believe that sun rays may be very useful for treating patients who have suffered from various illnesses.

These are some conclusions reached by researchers worldwide about the benefits of sun rays for human health:

- Conclusions reached in Ref. [68]: Regular sun exposure benefits health.
- Conclusions reached in Ref. [69]: Sunray exposure has a more positive effect on bone structure and homeostasis than vitamin D supplementation and control.
- Conclusions reached in Ref. [70]: A variety of medical claims were made for ultraviolet radiation, including that it increased resistance to diseases.
- Conclusions reached in Ref. [71]: There are relationships between sunlight and diseases. As a result, either too much or too little exposure to sun rays may be harmful to our health.
- Conclusions reached in Ref. [72]: This study's findings showed that regular sunlight exposure augments rates of COVID-19 recovery, correlating significantly and positively to COVID-19 patients' recoveries.

5.6 Conclusions

In this chapter, different ANN techniques/methods combined with IVS methods have been reviewed and then applied for predicting solar radiation intensity. Based on the information presented here, the main conclusions of this work are as follows: (1) By using techniques of artificial intelligence as optimizers and as objective function computing methods, we can achieve a correlation closer to 1 that indicates a perfect relationship between measured values and predicted values. The main critic to be noted for the existing machine learning techniques is that it requires data (measures) to produce data (predictions). Hence, we cannot apply these models to other locations/sites. (2) A detailed discussion on modeling and forecasting of instantaneous, hourly, daily, and/or monthly solar radiation has been presented. (3) Data from 35 different Moroccan locations were used for both training and testing purposes. (4) For daily and monthly solar radiation forecasting, an intelligent approach combined with ANNs was adopted and presented, then described and discussed in detail. According to the proposed approach, the best combinations of inputs for predicting DGSR and/or MAGSR were determined. 128 and 85 best combinations were found for DGSR and MAGSR, respectively. (5) A large review of statistical indicators used by researchers worldwide was done and then applied. The statistical analysis indicated almost perfect correlation and approximation between the predicted values by the ANNs models, and the measured data. In addition, 128 and 85 new linear relationships/parametric models between the found best combination and the considered outputs were developed and proposed. (6) Finally, as an application of the solar energy forecasting that we presented, a detailed review of solar energy conversion systems was also presented. This review was done for solar space heating and cooling, water heating, cooking, electrical generation, and several other conversion systems of sunlight into a useful energy form. Various applications were analyzed and discussed for systems powered by solar energy.

Appendix

Table 5.A1 Statistical analysis results for the best combinations of inputs in the forecasting of MAGSR.

Combination	R^2	MBE	RMSE	NRMSE	MAPE	MPE	TS	σ	SBF	LCE	WIA	ϕ	Rank
Comb52	0.983	3.788	9.314	0.055	1.585	−3.316	9.111	8.520	0.873	0.948	0.997	167	18
Comb67	0.983	3.784	9.311	0.055	1.583	−3.304	9.106	8.517	0.874	0.948	0.997	154	13
Comb70	0.983	3.770	9.318	0.056	1.582	−3.305	9.055	8.532	0.873	0.948	0.997	156	14
Comb73	0.986	4.333	8.758	0.052	1.369	−3.515	11.652	7.620	0.888	0.954	0.996	299	61
Comb200	0.986	4.315	8.703	0.052	1.366	−3.493	11.689	7.566	0.889	0.955	0.996	260	41
Comb202	0.982	4.018	9.454	0.056	1.628	−3.451	9.610	8.568	0.872	0.947	0.996	293	57
Comb207	0.982	4.140	9.615	0.057	1.661	−3.546	9.767	8.688	0.870	0.945	0.996	388	78
Comb216	0.982	3.997	9.465	0.056	1.628	−3.444	9.536	8.590	0.872	0.946	0.996	289	56
Comb220	0.987	4.271	8.686	0.052	1.352	−3.474	11.558	7.572	0.889	0.955	0.996	228	28
Comb223	0.986	4.307	8.772	0.052	1.361	−3.506	11.539	7.650	0.887	0.954	0.996	277	51
Comb226	0.982	3.962	9.424	0.056	1.619	−3.419	9.484	8.561	0.872	0.947	0.997	236	31
Comb235	0.983	3.775	9.297	0.055	1.585	−3.302	9.094	8.506	0.873	0.948	0.997	142	9
Comb238	0.986	4.282	8.719	0.052	1.355	−3.485	11.540	7.605	0.888	0.955	0.996	249	37
Comb245	0.983	3.811	9.255	0.055	1.583	−3.314	9.249	8.444	0.874	0.949	0.997	142	10
Comb312	0.986	4.324	8.785	0.052	1.367	−3.514	11.573	7.657	0.887	0.954	0.996	298	60
Comb313	0.986	4.352	8.758	0.052	1.374	−3.521	11.723	7.609	0.889	0.954	0.996	299	62
Comb314	0.982	4.136	9.616	0.057	1.659	−3.542	9.753	8.691	0.870	0.945	0.996	385	77
Comb315	0.986	4.364	8.834	0.053	1.376	−3.540	11.631	7.690	0.887	0.953	0.996	322	66
Comb317	0.983	3.862	9.303	0.055	1.594	−3.340	9.339	8.474	0.875	0.948	0.997	161	17
Comb321	0.982	3.986	9.472	0.056	1.628	−3.437	9.497	8.603	0.872	0.946	0.996	281	54
Comb322	0.988	3.785	8.044	0.048	1.288	−3.098	10.916	7.106	0.900	0.961	0.997	81	2
Comb329	0.983	3.966	9.384	0.056	1.617	−3.406	9.545	8.515	0.874	0.947	0.997	220	27
Comb332	0.983	3.952	9.388	0.056	1.617	−3.402	9.499	8.526	0.874	0.947	0.997	215	23
Comb334	0.985	4.157	8.913	0.053	1.498	−3.434	10.791	7.894	0.886	0.953	0.996	250	38

(Continued)

Table 5.A1 (Continued)

Combination	R^2	MBE	RMSE	NRMSE	MAPE	MPE	TS	σ	SBF	LCE	WIA	ϕ	Rank
Comb337	0.987	4.238	8.618	0.051	1.344	−3.446	11.563	7.512	0.890	0.956	0.996	208	22
Comb338	0.983	3.913	9.336	0.056	1.605	−3.371	9.450	8.486	0.875	0.948	0.997	180	19
Comb340	0.986	4.316	8.734	0.052	1.363	−3.502	11.637	7.602	0.888	0.954	0.996	276	48
Comb349	0.986	4.209	8.690	0.052	1.388	−3.427	11.334	7.611	0.890	0.955	0.996	219	26
Comb351	0.982	3.968	9.435	0.056	1.620	−3.424	9.488	8.571	0.872	0.947	0.997	253	39
Comb352	0.989	3.953	7.848	0.047	1.268	−3.201	11.934	6.788	0.902	0.963	0.997	136	8
Comb354	0.985	4.248	8.888	0.053	1.464	−3.468	11.138	7.817	0.889	0.953	0.996	255	40
Comb357	0.983	3.783	9.243	0.055	1.579	−3.293	9.183	8.443	0.875	0.949	0.997	123	5
Comb358	0.983	3.857	9.300	0.055	1.598	−3.340	9.331	8.472	0.874	0.948	0.997	158	15
Comb360	0.983	3.972	9.396	0.056	1.621	−3.423	9.549	8.525	0.873	0.947	0.997	234	30
Comb363	0.986	4.319	8.729	0.052	1.365	−3.502	11.654	7.595	0.888	0.954	0.996	278	52
Comb370	0.982	4.115	9.571	0.057	1.654	−3.525	9.748	8.652	0.871	0.945	0.996	364	75
Comb373	0.982	3.988	9.451	0.056	1.626	−3.435	9.528	8.578	0.872	0.947	0.996	272	45
Comb410	0.986	4.409	8.928	0.053	1.384	−3.579	11.624	7.773	0.886	0.952	0.996	334	70
Comb411	**0.987**	**3.438**	**8.087**	**0.048**	**1.294**	**−2.932**	**9.613**	**7.329**	**0.894**	**0.961**	**0.997**	**43**	**1**
Comb412	0.983	4.002	9.402	0.056	1.621	−3.425	9.628	8.518	0.874	0.947	0.996	247	35
Comb413	0.982	4.238	9.691	0.058	1.676	−3.615	9.954	8.725	0.869	0.944	0.996	431	85
Comb414	0.986	4.370	8.794	0.052	1.375	−3.536	11.721	7.641	0.888	0.954	0.996	319	65
Comb416	0.982	4.069	9.488	0.057	1.640	−3.483	9.719	8.581	0.872	0.946	0.996	324	67
Comb417	0.983	3.947	9.376	0.056	1.614	−3.394	9.499	8.515	0.874	0.947	0.997	202	21
Comb419	0.982	4.184	9.609	0.057	1.666	−3.561	9.901	8.660	0.871	0.945	0.996	396	80
Comb421	0.986	3.930	8.561	0.051	1.358	−3.240	10.575	7.615	0.890	0.956	0.997	133	7
Comb423	0.982	4.129	9.551	0.057	1.655	−3.528	9.814	8.622	0.871	0.946	0.996	363	74
Comb425	0.987	4.282	8.686	0.052	1.352	−3.478	11.600	7.565	0.889	0.955	0.996	231	29
Comb426	0.982	4.008	9.450	0.056	1.629	−3.442	9.585	8.568	0.873	0.947	0.996	278	53
Comb428	0.982	4.170	9.628	0.057	1.666	−3.564	9.837	8.688	0.870	0.945	0.996	400	81

Comb432	0.983	3.955	9.383	0.056	1.616	−3.400	9.513	8.520	0.874	0.947	0.997	215	24
Comb436	0.983	3.866	9.282	0.055	1.595	−3.337	9.378	8.449	0.875	0.949	0.997	159	16
Comb438	0.986	4.319	8.876	0.053	1.407	−3.526	11.400	7.764	0.886	0.953	0.996	301	63
Comb440	0.983	3.831	9.282	0.055	1.588	−3.328	9.275	8.464	0.874	0.949	0.997	151	12
Comb441	0.983	3.786	9.194	0.055	1.576	−3.285	9.249	8.388	0.876	0.950	0.997	118	4
Comb447	0.986	4.269	8.699	0.052	1.366	−3.475	11.529	7.588	0.888	0.955	0.996	237	33
Comb448	0.982	4.153	9.581	0.057	1.660	−3.541	9.845	8.645	0.871	0.945	0.996	379	76
Comb450	0.986	4.290	8.697	0.052	1.355	−3.484	11.608	7.574	0.889	0.955	0.996	248	36
Comb451	0.982	4.010	9.442	0.056	1.629	−3.444	9.602	8.559	0.873	0.947	0.996	276	49
Comb453	0.982	4.193	9.657	0.058	1.672	−3.581	9.868	8.709	0.869	0.944	0.996	418	83
Comb457	0.983	3.814	9.217	0.055	1.582	−3.299	9.304	8.401	0.876	0.949	0.997	129	6
Comb459	0.986	4.327	8.752	0.052	1.365	−3.511	11.643	7.616	0.888	0.954	0.996	294	58
Comb462	0.983	3.889	9.339	0.056	1.602	−3.368	9.374	8.501	0.873	0.948	0.997	185	20
Comb475	0.982	4.185	9.616	0.057	1.668	−3.563	9.896	8.668	0.871	0.945	0.996	403	82
Comb476	0.986	4.313	8.744	0.052	1.357	−3.504	11.606	7.615	0.888	0.954	0.996	275	47
Comb479	0.982	4.086	9.495	0.057	1.644	−3.491	9.758	8.581	0.872	0.946	0.996	331	69
Comb481	0.982	4.178	9.606	0.057	1.666	−3.555	9.887	8.660	0.871	0.945	0.996	388	79
Comb482	0.982	4.068	9.475	0.056	1.640	−3.480	9.732	8.567	0.872	0.946	0.996	313	64
Comb483	0.983	3.699	9.101	0.054	1.554	−3.225	9.106	8.325	0.877	0.951	0.997	101	3
Comb484	0.982	4.057	9.502	0.057	1.640	−3.487	9.666	8.602	0.871	0.946	0.996	325	68
Comb485	0.983	4.010	9.383	0.056	1.620	−3.430	9.676	8.493	0.874	0.947	0.996	242	34
Comb487	0.986	4.315	8.760	0.052	1.391	−3.498	11.585	7.633	0.889	0.954	0.996	276	50
Comb488	0.982	4.002	9.437	0.056	1.627	−3.437	9.586	8.556	0.873	0.947	0.996	266	43
Comb489	0.982	4.193	9.663	0.058	1.673	−3.581	9.859	8.716	0.869	0.944	0.996	419	84
Comb492	0.987	4.252	8.637	0.051	1.344	−3.455	11.576	7.527	0.889	0.955	0.996	217	25

(*Continued*)

Table 5.A1 (Continued)

Combination	R^2	MBE	RMSE	NRMSE	MAPE	MPE	TS	σ	SBF	LCE	WIA	ϕ	Rank
Comb493	0.983	3.994	9.431	0.056	1.626	−3.438	9.569	8.554	0.872	0.947	0.996	265	42
Comb496	0.986	4.303	8.775	0.052	1.432	−3.511	11.516	7.657	0.888	0.954	0.996	269	44
Comb497	0.983	3.850	9.254	0.055	1.590	−3.323	9.366	8.425	0.876	0.949	0.997	143	11
Comb499	0.982	4.088	9.531	0.057	1.648	−3.506	9.720	8.620	0.871	0.946	0.996	347	71
Comb503	0.983	4.030	9.431	0.056	1.631	−3.453	9.676	8.536	0.873	0.947	0.996	273	46
Comb504	0.982	4.091	9.546	0.057	1.649	−3.513	9.707	8.636	0.870	0.946	0.996	352	73
Comb505	0.987	4.284	8.688	0.052	1.352	−3.479	11.600	7.568	0.889	0.955	0.996	236	32
Comb506	0.983	4.044	9.443	0.056	1.634	−3.463	9.701	8.543	0.873	0.947	0.996	286	55
Comb507	0.983	4.138	9.428	0.056	1.629	−3.494	9.999	8.481	0.876	0.947	0.996	294	59
Comb509	0.982	4.114	9.527	0.057	1.650	−3.510	9.799	8.603	0.872	0.946	0.996	350	72

Note: Bold signifies the first best combination of inputs.

Table 5.A2 Linear formulas for the best combinations of inputs in the forecasting of MAGSR.

Combination	a0	a1	a2	a3	a4	a5	a6	a7	a8	a9	Rank
Comb52	−46.0563	0.9817	0	0	0	0	0	0	0.5467	76.4964	18
Comb67	−54.1137	0	0	0	2.2566	0	0	0	0.6231	46.6255	13
Comb70	−51.5780	0	0	0	0	0.8438	0	0	0.6210	76.7627	14
Comb73	−40.1823	0	0	0	0	0	0	−4.8494	0.6472	79.7665	61
Comb200	0.8261	0	0	0	0	0	−0.5057	−0.9533	0.6134	54.7928	41
Comb202	−1.4768	0	0	0	0	0.4719	−0.5193	0	0.6079	54.9311	57
Comb207	−50.9642	0	0	0	2.2015	−1.3994	0	0	0.6008	57.8049	78
Comb216	−60.9168	0	0	0.7138	2.2509	0	0	0	0.5693	67.8174	56
Comb220	−35.8628	0	1.2218	0	0	0	0	−1.4071	0.5324	58.8122	28
Comb223	−50.4972	0	1.3292	0	0	1.2878	0	0	0.5337	64.5515	51
Comb226	−80.1566	0	0.5917	0	1.4851	0	0	0	0.5800	99.7005	31
Comb235	−81.8498	1.1023	0	0	0	0	0	−0.8125	0.5490	135.6884	9
Comb238	−40.2176	1.0891	0	0	0	0.1316	0	0	0.5186	77.1813	37
Comb245	−43.7245	5.2110	0	−4.4134	0	0	0	0	0.5318	51.5848	10
Comb312	2.1425	0	0	0	0	0.1954	−0.5124	−0.9484	0.6144	52.1817	60
Comb313	9.0813	0	0	0	0.0959	0	−0.5298	−0.9459	0.6098	43.9159	62
Comb314	−52.1546	0	0	0	2.0070	−1.2796	0	−0.6127	0.6081	62.8847	77
Comb315	−3.4194	0	0	0	0.4501	0.2280	−0.4870	0	0.6128	45.9782	66
Comb317	11.1730	0	0	−0.1891	0	0	−0.5465	−0.9188	0.6250	40.4401	17
Comb321	−50.7733	0	0	0.8531	2.1319	0	0	−0.6029	0.5605	58.1962	54
Comb322	3.4860	0	0	−0.2344	0.2467	0	−0.5331	0	0.6174	46.5035	2
Comb329	14.2544	0	−0.0594	0	0	0.5957	−0.5685	0	0.6139	33.0954	27
Comb332	−1.1740	0	−0.1758	0	0.6918	0	−0.5008	0	0.6169	45.5064	23
Comb334	−60.0318	0	0.9648	0	0.8488	−0.4257	0	0	0.5421	82.2627	38

(Continued)

Table 5.A2 (Continued)

Combination	a0	a1	a2	a3	a4	a5	a6	a7	a8	a9	Rank
Comb337	−63.5671	0	2.1498	−1.4959	0	0	0	−0.8007	0.6114	59.4264	22
Comb338	9.0946	0	0.3001	−0.5367	0	0	−0.5386	0	0.6189	36.3260	19
Comb340	−50.3866	0	2.3233	−1.6196	0	−0.5759	0	0	0.5601	56.0305	48
Comb349	1.9275	−0.2035	0	0	0	0.2350	−0.5438	0	0.6191	53.3592	26
Comb351	−47.4754	0.8807	0	0	1.6448	0	0	−0.9307	0.5587	56.0670	39
Comb352	−3.9214	−0.1706	0	0	0.4502	0	−0.4978	0	0.6160	52.3693	8
Comb354	−57.9418	1.2824	0	0	2.0032	5.1505	0	0	0.5980	13.0747	40
Comb357	−47.8312	4.3947	0	−3.5853	0	0	0	−1.1476	0.5555	60.1315	5
Comb358	−3.3634	0.8540	0	−1.0204	0	0	−0.4925	0	0.6131	51.4595	15
Comb360	−40.9233	4.5768	0	−3.2123	0	5.5874	0	0	0.5532	10.4186	30
Comb363	−35.7883	22.3768	0	−21.7151	−8.5747	0	0	0	0.5165	51.9801	52
Comb370	−52.7854	−4.4816	5.0526	0	0	−1.6062	0	0	0.5803	47.7190	75
Comb373	−37.4149	20.5831	−19.8846	0	12.0127	0	0	0	0.5103	59.7177	45
Comb410	9.3631	0	0	0	0.4265	0.2064	−0.5014	−0.8285	0.6185	29.5620	70
Comb411	**15.0475**	**0**	**0**	**−0.2171**	**0**	**0.0905**	**−0.5520**	**−0.9096**	**0.6296**	**32.6290**	**1**
Comb412	0.0117	0	0	−0.1082	0.2122	0	−0.4990	−0.8062	0.6191	50.8097	35
Comb413	−32.8558	0	0	1.6610	2.2576	1.8168	0	−2.3317	0.5316	27.8914	85
Comb414	7.8129	0	0	−0.1888	0.5991	0.0670	−0.5068	0	0.6217	27.7110	65
Comb416	8.4302	0	−0.1839	0	0	−0.2092	−0.5564	−0.9962	0.6181	53.3710	67
Comb417	−1.3844	0	−0.1054	0	0.3134	0	−0.4953	−0.8012	0.6188	52.8446	21
Comb419	12.2882	0	−0.3174	0	0.4997	−0.3258	−0.5662	0	0.6191	38.5947	80
Comb421	8.5115	0	0.1264	−0.3020	0	0	−0.5323	−0.8202	0.6267	39.4945	7
Comb423	−3.5298	0	0.4604	−0.6415	0	−0.1122	−0.4896	0	0.6176	49.2365	74
Comb425	−52.4203	0	0.7187	0.5390	0.1797	0	0	−1.6213	0.4974	109.2216	29
Comb426	−2.3496	0	0.0741	−0.2565	0.3306	0	−0.4983	0	0.6173	48.8641	53
Comb428	−44.2944	0	1.0527	−0.3212	1.3739	−0.4891	0	0	0.5592	43.8757	81

Comb432	2.3940	−0.1127	0	0	0.2626	0	−0.5071	−0.8273	0.6192	48.0369	24
Comb436	12.5245	0.5527	0	−0.7491	0	0	−0.5349	−0.6564	0.6238	31.7790	16
Comb438	−2.2768	1.2137	0	−1.3517	0	−0.0257	−0.4749	0	0.6137	44.1934	63
Comb440	−55.9096	35.1321	0	−34.6800	−15.0015	0	0	0.1584	0.5094	95.3022	12
Comb441	−1.1509	6.4412	0	−6.6028	−2.7252	0	−0.4769	0	0.5983	53.3086	4
Comb447	−66.3783	−6.4834	7.2094	0	0	−0.3912	0	−0.1560	0.6216	24.2718	33
Comb448	12.0400	−0.9205	0.6480	0	0	−0.2040	−0.5694	0	0.6195	37.4133	76
Comb450	−25.3387	15.9471	−15.1003	0	9.4224	0	0	−0.8109	0.5163	39.9942	36
Comb451	−1.6073	6.1660	−6.3267	0	3.7318	0	−0.4765	0	0.5991	53.6885	49
Comb453	−36.8354	23.0319	−22.3513	0	13.2355	−0.1234	0	0	0.5041	63.3430	83
Comb457	3.2988	7.0627	−2.9691	−4.2682	0	0	−0.4824	0	0.5990	45.5477	6
Comb459	−22.3917	20.8610	−8.0888	−11.8761	0	1.4777	0	0	0.5055	29.4321	58
Comb462	−33.8532	25.5221	−11.7184	−13.1293	1.4109	0	0	0	0.4996	60.0531	20
Comb475	5.5155	0	−0.0778	0	0.1558	0.2375	−0.5292	−0.7366	0.6202	44.5634	82
Comb476	6.0880	0	0.9341	−0.8790	0	0.5445	−0.4519	−0.5634	0.6247	14.6253	47
Comb479	9.2835	0	0.0674	−0.2622	0.3296	0.5701	−0.5354	0	0.6233	29.2276	69
Comb481	10.5494	−0.1073	0	0	0.2846	0.2377	−0.5274	−0.8576	0.6228	33.8439	79
Comb482	13.6662	−1.2682	0	0.9189	0	−0.9165	−0.6253	−1.1171	0.6160	66.6994	64
Comb483	10.5992	−0.6449	0	0.5295	0.6752	0	−0.5182	−0.8412	0.6251	30.4588	3
Comb484	−41.4117	25.2349	0	−24.5457	−10.3494	0.1646	0	−0.4450	0.5069	73.7528	68
Comb485	1.1422	6.6595	0	−6.8255	−2.7943	−0.0141	−0.4794	0	0.5988	49.1291	34
Comb487	9.4471	−1.0266	0.9525	0	0	0.1778	−0.4981	−0.8228	0.6239	27.7301	50
Comb488	8.9324	−0.9596	0.8693	0	0.2102	0	−0.4905	−0.7636	0.6287	23.5212	43
Comb489	−29.8444	17.9841	−17.2349	0	10.7055	−0.1937	0	−0.6107	0.5206	44.4846	84
Comb492	11.0192	−1.1911	0.9158	0.1174	0	0	−0.5261	−0.8437	0.6283	30.4293	25

(Continued)

Table 5.A2 (Continued)

Combination	a0	a1	a2	a3	a4	a5	a6	a7	a8	a9	Rank
Comb493	−32.5637	26.6052	−10.8444	−15.0396	0	0.3171	0	−0.3718	0.5043	56.0484	42
Comb496	−27.3514	24.4650	−11.2288	−12.4613	1.2324	0	0	−0.6900	0.4959	56.6870	44
Comb497	1.7053	7.6305	−3.7005	−4.1019	0.4014	0	−0.4805	0	0.5961	50.2195	11
Comb499	−30.8731	27.9760	−12.9437	−14.3658	1.4221	−0.0562	0	0	0.4920	59.4745	71
Comb503	7.9180	0.0640	0.1547	−0.3667	0.5214	0.1237	−0.4972	0	0.6229	23.1164	46
Comb504	−16.2286	20.5934	−9.4648	−10.2918	0.8270	−1.2202	0	−1.3333	0.4780	62.5737	73
Comb505	11.2495	−0.3617	0.1808	−0.0101	0.1910	0	−0.5392	−0.8676	0.6262	35.5711	32
Comb506	9.7215	−1.1967	1.1060	−0.0009	0	0.1915	−0.5045	−0.7878	0.6279	25.0330	55
Comb507	3.0207	3.8561	−3.9525	0	2.3850	0.0672	−0.4857	−0.7337	0.6098	46.4578	59
Comb509	8.2019	0	−0.0049	−0.1532	0.1484	−0.1291	−0.5302	−0.8720	0.6207	42.8641	72

Note: Bold signifies the first best combination of inputs.

References

[1] M. Huber, R. Knutti, Anthropogenic and natural warming inferred from changes in Earth's energy balance, Nature Geoscience 5 (2012) 31–36. Available from: https://doi.org/10.1038/ngeo1327.

[2] J. Cook, N. Oreskes, P.T. Doran, W.R.L. Anderegg, B. Verheggen, E.W. Maibach, et al., Consensus on consensus: a synthesis of consensus estimates on human-caused global warming, Environmental Research Letters 11 (2016) 048002. Available from: https://doi.org/10.1088/1748-9326/11/4/048002.

[3] G.D.J. Harper, Solar energy projects for the evil genius, Evil Genius Series, first ed., McGraw-Hill, New York, 2007.

[4] W. Qin, L. Wang, A. Lin, M. Zhang, X. Xia, B. Hu, et al., Comparison of deterministic and data-driven models for solar radiation estimation in China, Renewable and Sustainable Energy Reviews 81 (2018) 579–594. Available from: https://doi.org/10.1016/j.rser.2017.08.037.

[5] M. Despotovic, V. Nedic, D. Despotovic, S. Cvetanovic, Review and statistical analysis of different global solar radiation sunshine models, Renewable and Sustainable Energy Reviews 52 (2015) 1869–1880. Available from: https://doi.org/10.1016/j.rser.2015.08.035.

[6] Z.A. Al-Mostafa, A.H. Maghrabi, S.M. Al-Shehri, Sunshine-based global radiation models: a review and case study, Energy Conversion and Management 84 (2014) 209–216. Available from: https://doi.org/10.1016/j.enconman.2014.04.021.

[7] G.E. Hassan, M.E. Youssef, Z.E. Mohamed, M.A. Ali, A.A. Hanafy, New temperature-based models for predicting global solar radiation, Applied Energy 179 (2016) 437–450. Available from: https://doi.org/10.1016/j.apenergy.2016.07.006.

[8] A. Castillejo-Cuberos, R. Escobar, Detection and characterization of cloud enhancement events for solar irradiance using a model-independent, statistically-driven approach, Solar Energy 209 (2020) 547–567. Available from: https://doi.org/10.1016/j.solener.2020.09.046.

[9] Almorox, J., Estimating Global Solar Radiation from Common Meteorological Data in Aranjuez, Spain 12, n.d.

[10] S. Kirmani, M. Jamil, M. Rizwan, Empirical correlation of estimating global solar radiation using meteorological parameters, International Journal of Sustainable Energy 34 (2015) 327–339. Available from: https://doi.org/10.1080/14786451.2013.826222.

[11] Y. El Mghouchi, T. Ajzoul, D. Taoukil, A. El Bouardi, The most suitable prediction model of the solar intensity, on horizontal plane, at various weather conditions in a specified location in Morocco, Renewable and Sustainable Energy Reviews 54 (2016) 84–98. Available from: https://doi.org/10.1016/j.rser.2015.09.089.

[12] N. Samuel Chukwujindu, A comprehensive review of empirical models for estimating global solar radiation in Africa, Renewable and Sustainable Energy Reviews 78 (2017) 955–995. Available from: https://doi.org/10.1016/j.rser.2017.04.101.

[13] T. Li, J. Yue, J.M. Russell, X. Zhang, Long-term trend and solar cycle in the middle atmosphere temperature revealed from merged HALOE and SABER datasets, Journal of Atmospheric and Solar-Terrestrial Physics 212 (2021) 105506. Available from: https://doi.org/10.1016/j.jastp.2020.105506.

[14] P. Neelamegam, V.A. Amirtham, Prediction of solar radiation for solar systems by using ANN models with different back propagation algorithms, Journal of Applied Research and Technology (2016) 14. Available from: https://doi.org/10.1016/j.jart.2016.05.001.

[15] C. Renno, F. Petito, A. Gatto, ANN model for predicting the direct normal irradiance and the global radiation for a solar application to a residential building, Journal of Cleaner Production 135 (2016) 1298–1316. Available from: https://doi.org/10.1016/j.jclepro.2016.07.049.

[16] E.F. Alsina, M. Bortolini, M. Gamberi, A. Regattieri, Artificial neural network optimisation for monthly average daily global solar radiation prediction, Energy Conversion and Management 120 (2016) 320–329. Available from: https://doi.org/10.1016/j.enconman.2016.04.101.

[17] S. Belaid, A. Mellit, Prediction of daily and mean monthly global solar radiation using support vector machine in an arid climate, Energy Conversion and Management 118 (2016) 105–118. Available from: https://doi.org/10.1016/j.enconman.2016.03.082.

[18] J.-L. Chen, G.-S. Li, S.-J. Wu, Assessing the potential of support vector machine for estimating daily solar radiation using sunshine duration, Energy Conversion and Management 75 (2013) 311–318. Available from: https://doi.org/10.1016/j.enconman.2013.06.034.

[19] A. Aybar-Ruiz, S. Jiménez-Fernández, L. Cornejo-Bueno, C. Casanova-Mateo, J. Sanz-Justo, P. Salvador-González, et al., A novel grouping genetic algorithm–extreme learning machine approach for global solar radiation prediction from numerical weather models inputs, Solar Energy 132 (2016) 129–142. Available from: https://doi.org/10.1016/j.solener.2016.03.015.

[20] W. Ji, K.C. Chee, Prediction of hourly solar radiation using a novel hybrid model of ARMA and TDNN, Solar Energy 85 (2011) 808–817. Available from: https://doi.org/10.1016/j.solener.2011.01.013.

[21] H. Sun, D. Yan, N. Zhao, J. Zhou, Empirical investigation on modeling solar radiation series with ARMA–GARCH models, Energy Conversion and Management 92 (2015) 385–395. Available from: https://doi.org/10.1016/j.enconman.2014.12.072.

[22] A.K. Yadav, S.S. Chandel, Solar radiation prediction using artificial neural network techniques: a review, Renewable and Sustainable Energy Reviews 33 (2014) 772–781. Available from: https://doi.org/10.1016/j.rser.2013.08.055.

[23] M. Almaraashi, Investigating the impact of feature selection on the prediction of solar radiation in different locations in Saudi Arabia, Applied Soft Computing 66 (2018) 250–263. Available from: https://doi.org/10.1016/j.asoc.2018.02.029.

[24] Y. Masoudi-Sobhanzadeh, H. Motieghader, A. Masoudi-Nejad, FeatureSelect: a software for feature selection based on machine learning approaches, BMC Bioinformatics 20 (2019) 170. Available from: https://doi.org/10.1186/s12859-019-2754-0.

[25] Y. Masoudi-Sobhanzadeh, H. Motieghader, World Competitive Contests (WCC) algorithm: a novel intelligent optimization algorithm for biological and non-biological problems, Informatics in Medicine Unlocked 3 (2016) 15–28. Available from: https://doi.org/10.1016/j.imu.2016.06.002.

[26] A. Husseinzadeh Kashan, League Championship Algorithm (LCA): an algorithm for global optimization inspired by sport championships, Applied Soft Computing 16 (2014) 171–200. Available from: https://doi.org/10.1016/j.asoc.2013.12.005.

[27] J.H. Holland, Searching nonlinear functions for high values, Applied Mathematics and Computation 32 (1989) 255–274. Available from: https://doi.org/10.1016/0096-3003(89)90096-9.

[28] R. Eberhart, J. Kennedy, A new optimizer using particle swarm theory, in: MHS'95. Proceedings of the Sixth International Symposium on Micro Machine and Human Science. Presented at the MHS'95, 1995, pp. 39–43. https://doi.org/10.1109/MHS.1995.494215

[29] M. Dorigo, M. Birattari, T. Stutzle, Ant colony optimization, IEEE Computational Intelligence Magazine 1 (2006) 28–39. Available from: https://doi.org/10.1109/MCI.2006.329691.

[30] E. Atashpaz-Gargari, C. Lucas, Imperialist competitive algorithm: an algorithm for optimization inspired by imperialistic competition, 2007 IEEE Congress on Evolutionary Computation. Presented at the 2007 IEEE Congress on Evolutionoary Computation, IEEE, 2007, pp. 4661–4667. Available from: https://doi.org/10.1109/CEC.2007.4425083.

[31] M.R. Meybodi, H. Beigy, New learning automata based algorithms for adaptation of backpropagation algorithm parameters, International Journal of Neural Systems 12 (2002) 45–67. Available from: https://doi.org/10.1142/S012906570200090X.
[32] V.K. Patel, V.J. Savsani, Heat transfer search (HTS): a novel optimization algorithm, Information Sciences 324 (2015) 217–246. Available from: https://doi.org/10.1016/j.ins.2015.06.044.
[33] M. Ghaemi, M.-R. Feizi-Derakhshi, Forest optimization algorithm, Expert Systems with Applications 41 (2014) 6676–6687. Available from: https://doi.org/10.1016/j.eswa.2014.05.009.
[34] A.E.-S. Ezugwu, A.O. Adewumi, Discrete symbiotic organisms search algorithm for travelling salesman problem, Expert Systems with Applications 87 (2017) 70–78. Available from: https://doi.org/10.1016/j.eswa.2017.06.007.
[35] R. Rajabioun, Cuckoo optimization algorithm, Applied Soft Computing 11 (2011) 5508–5518. Available from: https://doi.org/10.1016/j.asoc.2011.05.008.
[36] J.A. Sabbagh, A.A.M. Sayigh, E.M.A. El-Salam, Estimation of the total solar radiation from meteorological data, Solar Energy 19 (1977) 307–311. Available from: https://doi.org/10.1016/0038-092X(77)90075-5.
[37] H.O. Menges, C. Ertekin, M.H. Sonmete, Evaluation of global solar radiation models for Konya, Turkey, Energy Conversion and Management 47 (2006) 3149–3173. Available from: https://doi.org/10.1016/j.enconman.2006.02.015.
[38] J. Boland, B. Ridley, B. Brown, Models of diffuse solar radiation, Renewable Energy 33 (2008) 575–584. Available from: https://doi.org/10.1016/j.renene.2007.04.012.
[39] L.T. Wong, W.K. Chow, Solar radiation model, Applied Energy 69 (2001) 191–224. Available from: https://doi.org/10.1016/S0306-2619(01)00012-5.
[40] F.J. Batlles, M.A. Rubio, J. Tovar, F.J. Olmo, L. Alados-Arboledas, Empirical modeling of hourly direct irradiance by means of hourly global irradiance, Energy 25 (2000) 675–688. Available from: https://doi.org/10.1016/S0360-5442(00)00007-4.
[41] D. Rivas, S. Saleme-Vila, R. Ortega-Izaguirre, F. Chalé-Lara, F. Caballero-Briones, A climatological estimate of incident solar energy in Tamaulipas, northeastern Mexico, Renewable Energy 60 (2013) 293–301. Available from: https://doi.org/10.1016/j.renene.2013.05.022.
[42] P.I. Cooper, The absorption of radiation in solar stills, Solar Energy 12 (1969) 333–346. Available from: https://doi.org/10.1016/0038-092X(69)90047-4.
[43] V. Sivamadhavi, R.S. Selvaraj, Prediction of monthly mean daily global solar radiation using Artificial Neural Network, Journal of Earth System Science 121 (2012) 1501–1510. Available from: https://doi.org/10.1007/s12040-012-0235-1.
[44] A.K. Soteris, Solar Energy Engineering, Elsevier, 2014. Available from: https://doi.org/10.1016/C2011-0-07038-2.
[45] M. Iqbal, An Introduction to Solar Radiation, Elsevier, 1983. Available from: https://doi.org/10.1016/B978-0-12-373750-2.X5001-0.
[46] B.Y.H. Liu, R.C. Jordan, The interrelationship and characteristic distribution of direct, diffuse and total solar radiation, Solar Energy 4 (1960) 1–19. Available from: https://doi.org/10.1016/0038-092X(60)90062-1.
[47] A. Khosravi, R.O. Nunes, M.E.H. Assad, L. Machado, Comparison of artificial intelligence methods in estimation of daily global solar radiation, Journal of Cleaner Production 194 (2018) 342–358. Available from: https://doi.org/10.1016/j.jclepro.2018.05.147.
[48] C.G. Ozoegwu, Artificial neural network forecast of monthly mean daily global solar radiation of selected locations based on time series and month number, Journal of Cleaner Production 216 (2019) 1–13. Available from: https://doi.org/10.1016/j.jclepro.2019.01.096.

[49] B. Jamil, N. Akhtar, Comparative analysis of diffuse solar radiation models based on sky-clearness index and sunshine period for humid-subtropical climatic region of India: a case study, Renewable and Sustainable Energy Reviews 78 (2017) 329–355. Available from: https://doi.org/10.1016/j.rser.2017.04.073.

[50] Y. El Mghouchi, E. Chham, E.M. Zemmouri, A. El Bouardi, Assessment of different combinations of meteorological parameters for predicting daily global solar radiation using artificial neural networks, Building and Environment 149 (2019) 607–622. Available from: https://doi.org/10.1016/j.buildenv.2018.12.055.

[51] W. Xu, W. Chen, Y. Liang, Feasibility study on the least square method for fitting non-Gaussian noise data, Physica A: Statistical Mechanics and Its Applications 492 (2018) 1917–1930. Available from: https://doi.org/10.1016/j.physa.2017.11.108.

[52] H. Yuan, J. Zheng, L.L. Lai, Y.Y. Tang, A constrained least squares regression model, Information Sciences 429 (2018) 247–259. Available from: https://doi.org/10.1016/j.ins.2017.11.020.

[53] V. Badescu, Assessing the performance of solar radiation computing models and model selection procedures, Journal of Atmospheric and Solar-Terrestrial Physics 105–106 (2013) 119–134. Available from: https://doi.org/10.1016/j.jastp.2013.09.004.

[54] Y. El Mghouchi, T. Ajzoul, A. El Bouardi, Prediction of daily solar radiation intensity by day of the year in twenty-four cities of Morocco, Renewable and Sustainable Energy Reviews 53 (2016) 823–831. Available from: https://doi.org/10.1016/j.rser.2015.09.059.

[55] Y. El Mghouchi, A. El Bouardi, Z. Choulli, T. Ajzoul, New model to estimate and evaluate the solar radiation, International Journal of Sustainable Built Environment 3 (2014) 225–234. Available from: https://doi.org/10.1016/j.ijsbe.2014.11.001.

[56] S. Monjoly, M. André, R. Calif, T. Soubdhan, Hourly forecasting of global solar radiation based on multiscale decomposition methods: a hybrid approach, Energy 119 (2017) 288–298. Available from: https://doi.org/10.1016/j.energy.2016.11.061.

[57] S. Sun, S. Wang, G. Zhang, J. Zheng, A decomposition-clustering-ensemble learning approach for solar radiation forecasting, Solar Energy 163 (2018) 189–199. Available from: https://doi.org/10.1016/j.solener.2018.02.006.

[58] L.M. Halabi, S. Mekhilef, M. Hossain, Performance evaluation of hybrid adaptive neuro-fuzzy inference system models for predicting monthly global solar radiation, Applied Energy 213 (2018) 247–261. Available from: https://doi.org/10.1016/j.apenergy.2018.01.035.

[59] H. Jiang, Y. Dong, J. Wang, Y. Li, Intelligent optimization models based on hard-ridge penalty and RBF for forecasting global solar radiation, Energy Conversion and Management 95 (2015) 42–58. Available from: https://doi.org/10.1016/j.enconman.2015.02.020.

[60] T.R. Ayodele, A.S.O. Ogunjuyigbe, Prediction of monthly average global solar radiation based on statistical distribution of clearness index, Energy 90 (2015) 1733–1742. Available from: https://doi.org/10.1016/j.energy.2015.06.137.

[61] O.N. Mensour, B. El Ghazzani, B. Hlimi, A. Ihlal, Modeling of solar energy potential in Souss-Massa area-Morocco, using intelligence Artificial Neural Networks (ANNs), Energy Procedia, Materials & Energy I 139 (2015) (2017) 778–784. Available from: https://doi.org/10.1016/j.egypro.2017.11.287.

[62] R. Eke, A. Senturk, Performance comparison of a double-axis sun tracking vs fixed PV system, Solar Energy 86 (2012) 2665–2672. Available from: https://doi.org/10.1016/j.solener.2012.06.006.

[63] N.H. Reich, W.G.J.H.M. van Sark, W.C. Turkenburg, W.C. Sinke, Using CAD software to simulate PV energy yield – the case of product integrated photovoltaic operated under indoor solar irradiation, Solar Energy 84 (2010) 1526–1537. Available from: https://doi.org/10.1016/j.solener.2010.05.015.

[64] A.E. Ghitas, Studying the effect of spectral variations intensity of the incident solar radiation on the Si solar cells performance, NRIAG Journal of Astronomy and Geophysics 1 (2012) 165–171. Available from: https://doi.org/10.1016/j.nrjag.2012.12.013.
[65] S. Kalogirou, The potential of solar industrial process heat applications, Applied Energy 76 (2003) 337–361. Available from: https://doi.org/10.1016/S0306-2619(02)00176-9.
[66] S. Bent, Solar Energy Storage, Elsevier, 2015. Available from: https://doi.org/10.1016/C2011-0-07146-6.
[67] B. Paul, Solar Power Generation, Elsevier, 2016. Available from: https://doi.org/10.1016/C2014-0-04849-6.
[68] H.J. van der Rhee, E. de Vries, J.W. Coebergh, Regular sun exposure benefits health, Medical Hypotheses 97 (2016) 34–37. Available from: https://doi.org/10.1016/j.mehy.2016.10.011.
[69] M.M.A. Abulmeaty, Sunlight exposure vs. vitamin D supplementation on bone homeostasis of vitamin D deficient rats, Clinical Nutrition Experimental 11 (2017) 1–9. Available from: https://doi.org/10.1016/j.yclnex.2016.10.003.
[70] M.R. Albert, K.G. Ostheimer, The evolution of current medical and popular attitudes toward ultraviolet light exposure: Part 2, Journal of the American Academy of Dermatology 48 (2003) 909–918. Available from: https://doi.org/10.1067/mjd.2003.272.
[71] N. Schüz, M. Eid, Sun exposure and skin cancer prevention, in: J.D. Wright (Ed.), International Encyclopedia of the Social & Behavioral Sciences, second ed., Elsevier, Oxford, 2015, pp. 696–700. Available from: https://doi.org/10.1016/B978-0-08-097086-8.14136-4.
[72] A. Asyary, M. Veruswati, Sunlight exposure increased Covid-19 recovery rates: a study in the central pandemic area of Indonesia, The Science of the Total Environment 729 (2020) 139016. Available from: https://doi.org/10.1016/j.scitotenv.2020.139016.

CHAPTER SIX

Digital twin predictive maintenance strategy based on machine learning improving facility management in built environment

S. Agostinelli[1] and A. Heydari[2]
[1]Department of Astronautics, Electrical and Energy Engineering, Sapienza University of Rome, Rome, Italy
[2]CITERA Interdepartmental Centre, Sapienza University of Rome, Rome, Italy

Contents

6.1 Introduction

Due to the widespread use of industrial machinery, a huge amount of data can be collected every day, resulting in increased attention of production managers and data analysts to potential applications. Predictive maintenance (PdM) is a prominent approach in this regard, which intends to monitor and analyze a system in real time to promptly identify and prevent potential maintenance needs [1].

Predictive maintenance (PdM) intends to prevent incipient failures by paying attention to unusual behaviors, providing a just-in-time maintenance intervention; such interventions are necessary to ensure both the quality and safety of devices as well as preventing unnecessary costs [2]. Building maintenance is

Artificial Neural Networks for Renewable Energy Systems and Real-World Applications.
DOI: https://doi.org/10.1016/B978-0-12-820793-2.00007-0

a crucial section of facility management (FM), mainly because maintenance-related expenditures account for at least 65% of FM costs each year [3].

Several studies have investigated PdM models. For instance, Quatrini et al. performed a wide literature review in this field and mentioned four main areas: (1) PdM fundamentals and implementation, (2) PdM strategies, (3) inspections and replacement plan, and (4) prognosis [4]. Many studies focused on PdM plans reported the remaining useful life (RUL) as a condition index. In addition, Zhou et al. reported a reliability-oriented model based on a PdM continuous monitoring system, which is based on the hypothesis that such systems are prone to continuous degradation that can be monitored [5]. Such degradation processes can be modeled using the Markov decision process [6], proportional hazards model [7], a gamma process [8], and Monte Carlo simulation [9,10]. Berka and Macek presented a model for fault identification to perform effective maintenance interventions. Susto et al. reported a PdM system based on the availability of the current values of the physical factors acting on the production process and on support vector machines (SVMs) [11]. In order to detect faulty and nonfaulty states of the machines, SVMs are applied, which can estimate the distance from failure as the equipment RUL. Yiwei et al. developed a prognostic model that can link the extended Kalman filter with a first-order perturbation technique. They developed a cost-based predictive maintenance (CDPM) framework and reported their model as a case study [12]. Wang et al. developed a cloud-based paradigm in order to predict maintenance using a mobile agent to allow in-time access to information and to apply them in order to enhance the accuracy and reliability of the fault identification process, estimating service life, and schedule maintenance procedures [13]. Nevertheless, their model does not contain a valid prediction algorithm for condition prediction. In another study, Ren and Zhao introduced a framework based on the Internet of Things (IoT) for both research and manufacturing, and an operation and maintenance (O&M) data obtaining procedure, based on a decision-making technique to predict the need for maintenance, including decision tree, k-means, support vector machine (SVM), and neural network [14]. Nevertheless, their study does not explain how to apply this technique and the prediction procedure. Cheng et al. performed a study to compare marker-based augmented reality (AR) and marker-less AR for indoor maintenance and decoration [15]. Nevertheless, this model is solely about maintenance and operational information and does not contain maintenance planning. Kang and Hong developed a software intended to effectively incorporate building information modeling (BIM) with a geographic information system (GIS)-based facilities management (FM) system [16]. In addition, they reported a BIM/GIS-based prototype intended to extract, transform, and load data from BIM and GIS to integrate FM data. Ayvaz and Alpay developed a maintenance system based on available data in order to predict production lines using machine learning techniques and the IoT [17]. Vafaei et al. presented an alarm system based on fuzzy logic to predict early equipment degradation in a car manufacturing line, with emphasis on minimizing costs related to sudden failures [18]. Wei et al. [5] introduced a conditional maintenance framework intended to identify the optimal action in order to reduce the mean cost rate as much as possible [19]. Cheng et al. [8] introduced a data-based model to predict maintenance of mechanical, electrical, and plumbing components based on information modeling and the IoT [20].

Based on what was mentioned before, the research gap contains (1) inadequate data integration for predictive maintenance, (2) absence of valid predictive patterns, and (3) no description of predictive procedures. The main novelties and contributions of this chapter are providing an intelligent predictive maintenance strategy for mechanical, electrical, and plumbing sections of building facilities, such as heating, ventilation and air conditioning (HVAC) systems, electrical parts, and lighting, that have a crucial role in ensuring the functionality of buildings.

6.2 Case study

The Rione Rinascimento complex hosts 3000 tenants in 950 apartments divided into seven building complexes. Each complex consists of three eight-floor buildings. It can be defined as an eco-neighborhood as it is powered by 65% renewable sources and 100% only regarding the thermal consumption for the production of hot water and air conditioning. This result has been achieved thanks to the high efficiency of the building envelope and the installation of the largest European geothermal power plant for residential use working through 190 geothermal probes at 150 m depth allowing favorable heat exchange with the ground both in winter and in summer conditions. In addition, a biomass cogeneration plant coupled with the geothermal plant allows economic savings of 40% to be reached as well as a decrease in CO_2 emissions of 50%.

The case study is significant due to peculiar conditions related to energy and safety management, as well as for building maintenance strategies. In fact, from the energy management point of view, it is considered as self-sufficient; considering the safety management, it is characterized as a well-defined system, physically separated from the surrounding urban context; from the maintenance point of view, it allows to obtain numerous and structured data as it is based on global services contract. This chapter investigates the maintenance aspect with a view to the optimization of planned maintenance processes through predictive systems.

6.2.1 Maintenance system

The maintenance datasheets are the starting point of the entire process, as they provide the necessary information about the analysis of the considered elements divided into categories/disciplines. In the case study, customized datasheets have been created for the mechanical and electrical system. Both are only considered in the internal and external common areas, thus excluding spaces for private use (Fig. 6.1).

The used maintenance datasheets only consider activities and elements that can be expected to cause faults and malfunctions, involving the replacement of single components. These sheets are regularly compiled for each scheduled maintenance operation.

6.3 Proposed predictive maintenance strategy

The PdM contains three key stages [21], including data acquisition: obtaining data from different sources that are equally important for PdM.

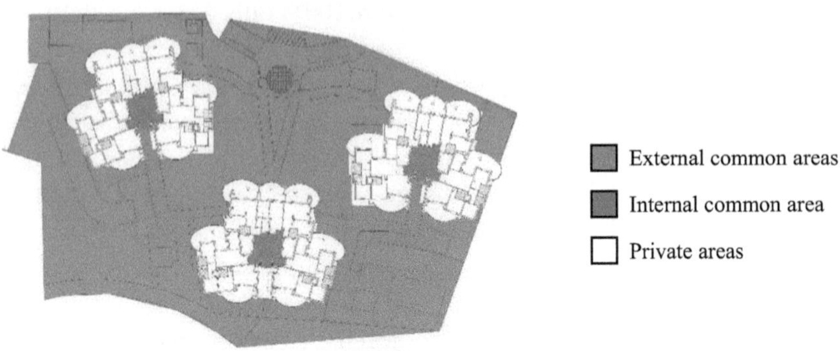

Figure 6.1 Building module in the Rione Rinascimento complex.

As it is structured, the data acquisition process is fully automated. The second is unstructured, and the collection can be partially entrusted to the operator; hence, accurate validation is required. Therefore data processing should be performed, because this process is crucial for dataset cleaning, as well as to analyze data to assess their consistency with the physical phenomena. The last process is the maintenance decision-making stage, which consists of two main groups: (1) fault identification and (2) fault prediction.

In this study, a maintenance strategy is presented for mechanical, electrical, and lighting systems supplying the external and common areas, excluding the apartments to avoid data perturbation due to different users' behaviors. The proposed strategy is implemented as follows:

1. Collecting data;
2. Determining the parameters affecting the system outputs;
3. Determine the optimal values of the obtained parameter;
4. Fault identification model;
5. Fault prediction intelligent strategy for electrical and mechanical systems.

In order to optimize the timing of the maintenance operations, the study started by considering the maximum initial values of the maintenance intervals taken from the Italian and European technical standards. A threshold value of tolerability of faults detected at the time of the maintenance visit was chosen, estimated at around 5% for all types of implemented components.

This choice is related to factors such as: avoiding excessively high malfunctioning values that would compromise the functionality of the systems, or at least penalize the quality of service; on the other hand, preventing an excessive increase in costs associated with scheduled maintenance activities.

The proposed methodology is based on a statistical evaluation of the deviation between the failure rate of a single device and the set threshold value (as mentioned above, 5% was chosen for uniformity and simplicity in this case study).

If this value is lower than the threshold, subsequent maintenance will be scheduled according to the time interval already provided by the standard; on the other hand, if the value of the failure rate is higher than the threshold limit, the scheduled time interval will be shortened, and subsequent maintenance will be anticipated. The shortened period will follow the percentage value exceeding the 5% threshold according to Eq. (6.1):

$$Mp = Mppl - [Mppl * (fr - frth)] \tag{6.1}$$

where *fr* is the real failure rate; *frth* is the threshold limit of the failure rate; *Mppl* is the planned maintenance period; and *Mp* is the shortened maintenance period;

Whenever a scheduled maintenance cycle of components is completed, the failure rate statistics must be evaluated and the timing for the next maintenance schedule is updated. When the failure rate curve falls below 5%, this means that the chosen interval is adequate and does not need to be shortened further. If particularly strong requirements for continuity of service of the installations are needed, such as for some essential components of the electrical power supply system, it will be sufficient to lower the threshold limit to reach the desired reliability values, varying the time interval of the maintenance operations accordingly.

6.4 Results and discussions

Therefore, for the application of the previously illustrated methodology, the components described in Table 6.1 with their base maintenance intervals obtained from the related regulations were taken into account in the period 2018–20.

By graphically describing the 3-year data obtained following the maintenance datasheets as described in Section 6.2.1, it is possible to evaluate the trends in the failure rate (shown on the y-axis) as a function of the shortening of the maintenance intervals carried out according to the formula described above.

Fig. 6.2 shows how the failure rate tends to fall below 5% after a period varying according to the performed maintenance cycles.

Table 6.1 Base maintenance frequency plan.

HVAC	Frequency	Electrical	Frequency
Boiler	Annual	Electrical panel	Monthly
Piping insulation	Annual		
Cooling system	Annual	Lighting systems lamps	Quarterly
AHU belt and motion parts	Monthly		
AHU flat filters	Quarterly	Lighting system equipment	Monthly
Valves	Monthly		

HVAC, heating, ventilation and air conditioning

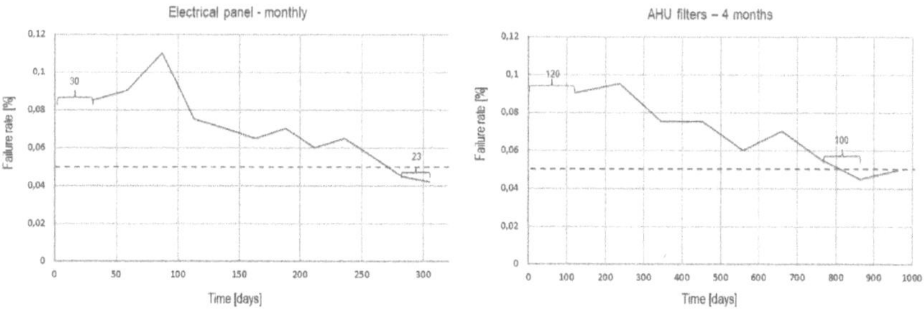

Figure 6.2 Electrical panel and air handling unit monthly and quarterly failure rate reduction.

For monthly maintenance activities, the threshold is reached after less than a year, while quarterly maintenance activities take about two and a half years; however, if the number of maintenance cycles is assessed [12 for the electrical panel and 9 for air handling unit (AHU) filters, respectively], it can be observed that the trend is similar.

The percentages of shortening of maintenance intervals are also similar and range from 24% for the electrical panel to 17% for AHU.

Therefore, it is always possible to reach the desired failure rate level with a proportional increase in expense, as maintenance intervals are shortened. Therefore, in order to optimize the operational cost–benefit, it is crucial to precisely identify the value of the most suitable threshold limit for the specific conditions.

6.4.1 Digital twin for building management systems

The proposed building management system can achieve considerably more efficiency by configuring the digital twin of the building through the integration of the BIM model aimed at the AIM (asset information model) [22] with information systems, building management systems, the IoT, machine learning, and mixed reality for the optimization and automation of maintenance activities.

In addition, working on the BIM model as the core of the management system architecture, it is possible to improve the potential related to physical/spatial information for space management, optimizing maintenance activities, and integrating machine learning systems and rule-based methods such as association rule mining [23].

Therefore, it is useful to create a hierarchy in the classification of spaces, using machine learning techniques defined as "clustering" automatically identifying groups/classes of similar spaces for their digital representation.

The proposed system therefore involves the integration of three components:

- Data related to BIM model objects (Autodesk Revit);
- Data flows programmed through visual programming systems (Autodesk Dynamo) relating bidirectionally to the Artificial intelligence (AI) systems with the BIM model;
- AI algorithm improving efficiency and compatibility with flows programmed through Dynamo [24]. The AI system can use mixed techniques of regression and hybrid reasoning.

6.5 Conclusions

Big Data is becoming a preponderant issue in many technological fields; predictive maintenance should develop innovative methodologies able to increase technical plants' reliability levels toward a near-zero failure rate.

In this chapter the applied methodologies show the importance of implementing Big Data unsupervised machine learning systems able to highlight connections between different maintenance conditions and strategies in order to obtain the most suitable threshold failure rate values for each kind of equipment.

Such an approach is confirmed by the failure of traditional supervised analyses, due to great difficulties in relating involved variables to their real impact on the maintenance strategies.

The collection of a great amount of maintenance data from different building systems, coupled with a data acquisition tool able to filter inadequate information will improve in a decisive way the predictive maintenance strategies in public and private organizations. In future research scenarios, progresses in cognitive technologies, artificial intelligence and calculation will provide enhanced capabilities to achieve a full-digital self-learned operation and maintenance approach and further developments will be focused on technical issues of the application on building and infrastructures.

References

[1] S. Selcuk, Predictive maintenance, its implementation and latest trends, Proceedings of the Institution of Mechanical Engineers, Part B: The Journal of Engineering Manufacture 231 (9) (2017) 1670–1679.

[2] L. Krishnamurthy, et al., Design and deployment of industrial sensor networks: experiences from a semiconductor plant and the North Sea, Proceedings of the 3rd International Conference on Embedded Networked Sensor Systems, Association for Computing Machinery, 2005, pp. 64–75.
[3] C. Eastman, P. Teicholz, R. Sacks, K. Liston, BIM Handbook: A Guide to Building Information Modeling for Owners, Managers, Designers, Engineers and Contractors, John Wiley & Sons, 2011.
[4] E. Quatrini, F. Costantino, G. Di Gravio, R. Patriarca, Condition-based maintenance—an extensive literature review, Machines 8 (2) (2020) 31.
[5] X. Zhou, L. Xi, J. Lee, Reliability-centered predictive maintenance scheduling for a continuously monitored system subject to degradation, Reliability Engineering and System Safety 92 (2007) 530–534. Available from: https://doi.org/10.1016/j.ress.2006.01.006.
[6] S. Sheu, C. Chang, Y. Chen, Z. George, Optimal preventive maintenance and repair policies for multi-state systems, Reliability Engineering and System Safety 140 (2015) 78–87. Available from: https://doi.org/10.1016/j.ress.2015.03.029.
[7] Z. Tian, H. Liao, Condition based maintenance optimization for multi-component systems using proportional hazards model, Reliability Engineering and System Safety 96 (5) (2011) 581–589. Available from: https://doi.org/10.1016/j.ress.2010.12.023.
[8] H. Liao, E.A. Elsayed, L. Chan, Maintenance of continuously monitored degrading systems, European Journal of Operational Research 175 (2006) 821–835. Available from: https://doi.org/10.1016/j.ejor.2005.05.017.
[9] M. Marseguerra, E. Zio, L. Podofillini, Condition-based maintenance optimization by means of genetic algorithms and Monte Carlo simulation, Reliability Engineering and System Safety 77 (2002) 151–165.
[10] J. Barata, C.G. Soares, M. Marseguerra, E. Zio, Simulation modelling of repairable multi-component deteriorating systems for 'on condition' maintenance optimisation, Reliability Engineering and System Safety 76 (2002) 255–264.
[11] G.A. Susto, A. Schirru, S. Pampuri, S. Mcloone, S. Member, A. Beghi, Machine learning for predictive maintenance: a multiple classifier approach, IEEE Transactions on Industrial Informatics 11 (3) (2015) 812–820.
[12] Y. Wang, C. Gogu, N. Binaud, C. Bes, R.T. Haftka, N.H. Kim, A cost driven predictive maintenance policy for structural airframe maintenance, Chinese Journal of Aeronautics 30 (3) (2017) 1242–1257. Available from: https://doi.org/10.1016/j.cja.2017.02.005.
[13] J. Wang, L. Zhang, L. Duan, R.X. Gao, A new paradigm of cloud-based predictive maintenance for intelligent manufacturing, Journal of Intelligent Manufacturing 28 (5) (2017) 1125–1137.
[14] S. Ren, X. Zhao, A predictive maintenance method for products based on big data analysis, Proceedings International Conference on Materials Engineering and Information Technology Applications (MEITA) (2015) 385–390.
[15] Comparison of marker-based AR and marker-less AR: a case study on indoor decoration system, in Lean and Computing in Construction Congress (LC3): Proceedings of the Joint Conference on Computing in Construction (JC3), 2017, pp. 483–490.
[16] T.W. Kang, C.H. Hong, Automation in Construction A study on software architecture for effective BIM / GIS-based facility management data integration, Automation in Construction 54 (2015) 25–38. Available from: https://doi.org/10.1016/j.autcon.2015.03.019.
[17] S. Ayvaz, K. Alpay, Predictive maintenance system for production lines in manufacturing: a machine learning approach using IoT data in real-time, Expert Systems with Applications 173 (2021) 114598. Available from: https://doi.org/10.1016/j.eswa.2021.114598. September 2020.

[18] N. Vafaei, R.A. Ribeiro, L.M. Camarinha-matos, Fuzzy early warning systems for condition based maintenance, Computers & Industrial Engineering 128 (2019) 736–746. Available from: https://doi.org/10.1016/j.cie.2018.12.056. December 2018.
[19] G. Wei, X. Zhao, S. He, Z. He, Reliability modeling with condition-based maintenance for binary-state deteriorating systems considering zoned shock effects, Computers & Industrial Engineering 130 (2019) 282–297. Available from: https://doi.org/10.1016/j.cie.2019.02.034. July 2018.
[20] J.C.P. Cheng, W. Chen, K. Chen, Q. Wang, Data-driven predictive maintenance planning framework for MEP components based on BIM and IoT using machine learning algorithms, Automation in Construction 112 (2020) 103087. Available from: https://doi.org/10.1016/j.autcon.2020.103087.
[21] A.K.S. Jardine, D. Lin, D. Banjevic, A review on machinery diagnostics and prognostics implementing condition-based maintenance, Mechanical Systems and Signal Processing 20 (2006) 1483–1510. Available from: https://doi.org/10.1016/j.ymssp.2005.09.012.
[22] UNI EN ISO 19650 Standard Organization and digitization of information about buildings and civil engineering works, including building information modelling (BIM)—Information management using building information modelling—Part 3: Operational phase of the assets.
[23] A. Telikani, A.H. Gandomi, A. Shahbahrami, A survey of evolutionary computation for association rule mining, Information Sciences 524 (2020) 318–352. Available from: https://doi.org/10.1016/j.ins.2020.02.073. ISSN 0020-0255.
[24] M. Salamak, M. Jasiński, T. Płaszczyk, M. Żarski, Analytical modelling in dynamo, Transactions of Vsb, Technical University of Ostrava, Civil Engineering Series 18 (2) (2018) 36–43. Available from: https://doi.org/10.31490/tces-2018-0014.

CHAPTER SEVEN

Artificial neural network and desalination systems

Fadl A. Essa[1], Mohamed Elasyed Abd Elaziz[2], S. Shanmugan[3] and Ammar H. Elsheikh[4]

[1]Mechanical Engineering Department, Faculty of Engineering, Kafrelsheikh University, Kafrelsheikh, Egypt
[2]Department of Mathematics, Faculty of Science, Zagazig University, Zagazig, Egypt
[3]Research Centre for Solar Energy, Department of Engineering Physics, College of Engineering, Koneru Lakshmaiah Education Foundation, Guntur, Andra Pradesh, India
[4]Production Engineering and Mechanical Design Department, Tanta University, Tanta, Egypt

Contents

Artificial Neural Networks for Renewable Energy Systems and Real-World Applications.
DOI: https://doi.org/10.1016/B978-0-12-820793-2.00010-0

7.1 Introduction

Water is one of the most important components of Earth, and it is essential for human life. It is available on Earth in abundance, but much less of it is available for human use. Fresh and potable water is the most prominent issue facing mankind at present. About 71% of the Earth is covered with water, of which 96.5% is ocean water and the rest is in rivers and lakes, ponds, ice caps and glaciers, and in the soil and aquifers, etc. From all of these, less than 1% of water is fresh and suitable for human consumption. The issue of potable water grows daily [1]. There are many factors responsible for the depletion of freshwater. Some of these include increasing population, industrialization, urbanization, transportation, etc. There is a need for water for various purposes such as cooking, farming, drinking, and many more. Thus, safe water is a big challenge for current and future generations [2].

The lack of access to freshwater has an adverse effect on the lives of ordinary people. There are many waterborne diseases which are spread only because of a lack of freshwater. Poor people are the primary victims of this crucial problem [3]. Also, in some of regions like deserts and arid regions, there is much lower rainfall, which causes adverse effects on human quality of life. the estimated global water scarcity for 2030 is shown in Fig. 7.1, which is based upon the Falkenmark Indicator.

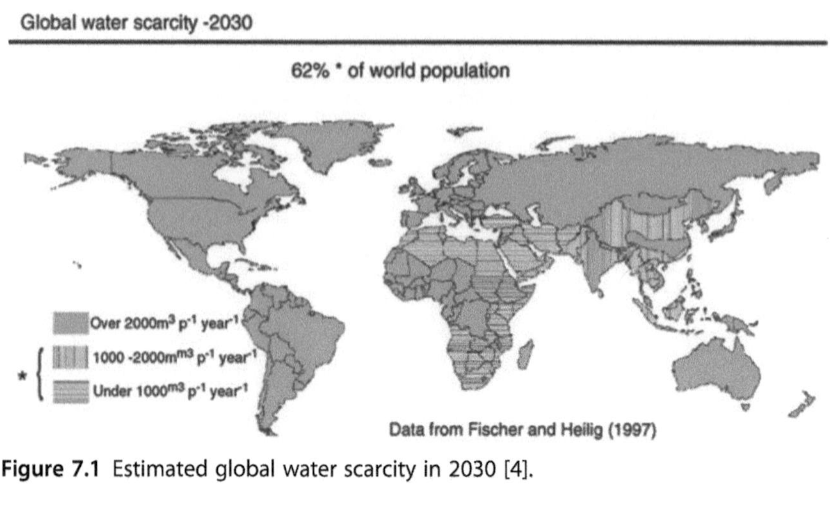

Figure 7.1 Estimated global water scarcity in 2030 [4].

Water occurs in a very complicated dynamic cycle that includes rain, evaporation, runoff, and many more dynamic natural processes, unlike land which is considered to be a static resource. Water controls nature and its components—in other words it manages the ecosystem. Sustainable water management is one of the biggest challenges of this era [5]. The main causes of this problem are the uneven distributions of both human population and water resources. It has been observed that densely populated regions have less water availability and lower populated regions generally have adequate or even surplus water availability.

It is not only physical water scarcity which creates problems for humankind but also social water scarcity, which is growing daily. Social water scarcity is a water issue which comes into play due to political parties, policies, and socioeconomic relationships [6].

Due to rapid population growth, the freshwater need increasing daily, and it is becoming increasingly difficult to fulfill everyone's freshwater needs [7]. A shortage of freshwater is damaging to the ecosystem, which is one of the most dangerous threats to humankind. Freshwater depletion is not an issue of arid regions only, rather it has become common throughout the world [8]. It is essential to keep control over the depletion of freshwater and to generate more fresh and potable water from other sources which are not useful to humans directly.

The principle of the hydrological cycle is followed in the manmade desalination process using other sources of heating and cooling. A large amount of energy is needed to separate freshwater from brine and salty seawater. Desalination takes place by feeding saltwater into the system which gives two output streams as a result, one is the freshwater stream and the other is the salt-contaminated water stream. Thus, freshwater is obtained by desalinating saltwater [9–13].

The desalination process has become a major method to supply freshwater in many regions of the world. The desalination process is mostly used in coastal regions as there is an abundant water source. The most important characteristic of this process is that it is safe for all—in other words it has no adverse effects on the ecosystem [14]. As per a survey carried out in the previous decade, about 75 million people all over the world are dependent on desalination processes to obtain freshwater for their daily needs. There are many countries which are dependent on desalination to obtain freshwater. The top five leading nations in respect of desalination plant capacity are Saudi Arabia, the United States, the UAE, Spain, and Kuwait, with percentages of 17.4, 16.2, 14.7, 6.4, and 5.8%,

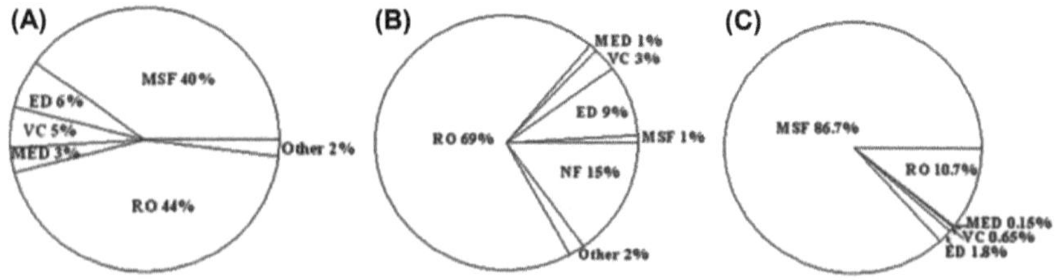

Figure 7.2 Desalination production capacity with different categories for (A) the world, (B) the United States, and (C) Middle Eastern countries in 2002 [16].

respectively [15]. The desalination production capacity for different nations is shown in Fig. 7.2.

There were about 18,000 desalination plants all over the world in 2015, with a total installed production capacity of 86.55 million m^3/day or 22,870 million gallons per day (MGD). Of the whole capacity, around 44% (37.32 million m^3/day or 9860 MGD) is located in the Middle East and North Africa. Due to current advancements in technologies related to desalination, the energy used for water production has been reduced by 80% over 20 years [16].

7.2 Methods of desalination

There are various methods of desalinating brackish and salty seawater. Commercially and economically, of all methods, multistage flash distillation (MSFD), reverse osmosis (RO), and multiple-effect distillation (MED) are most used for desalination. It has been observed that these three methods are the leading ones and in the future these three will be the most competitive [17]. There are various methods of desalination as described next.

7.2.1 Multistage flash distillation

The basic principle of MSFD is flash evaporation. In this process, evaporation of seawater takes place through the reduction of pressure as opposed to an increase in temperature. To get the maximum product and to maintain the economies of MSFD, regenerative heating generally is done. Due to regeneration, seawater flashing in the flashing chamber provides heat to the seawater going through the flashing method. As this is regenerative heating, this process needs different stages for completion. There is a need to raise the temperature of incoming seawater at each stage gradually [18]. This gradual increase in the temperature of the seawater is achieved through the heat of condensation, which is released by the condensing water vapor. There are basically three parameters essential for an MSFD plant, and these are heat input, heat recovery, and finally heat rejection. There is the possibility of scale formation in an MSFD plant, and to avoid that scale formation, high-temperature additives are used [19].

In a modern MSFD plant, multistage evaporators are used in which there are about 19–28 stages [20]. The operating temperature of an MSFD plant is in the range of 90–120 °C. As the operating temperature of the plant increases, there is an increase in the plant's efficiency, however this may lead to more scale formation. There is a need to maintain pressure below the corresponding saturation temperature of the heated seawater.

There are different types of equipment and accessories used in these plant for different purposes. These include demisters, decarbonators, and vacuum deaerators, as illustrated in Fig. 7.3. Demisters are provided at each stage of the evaporator to minimize carryover of brine droplets into the distillate. The purpose of using a decarbonator and vacuum deaerator is to remove dissolved gases from the brine [21].

The first MSFD plant was built in the 1950s [23]. The Saline Water Conversion Corporation's Al-Jubail plant in Saudi Arabia is the world's largest plant, with a capacity of around 815,120 m^3/day [24]. The largest MSF unit, with a capacity of 75,700 m^3/day, is the Shuweihat plant, situated in the UAE [25].

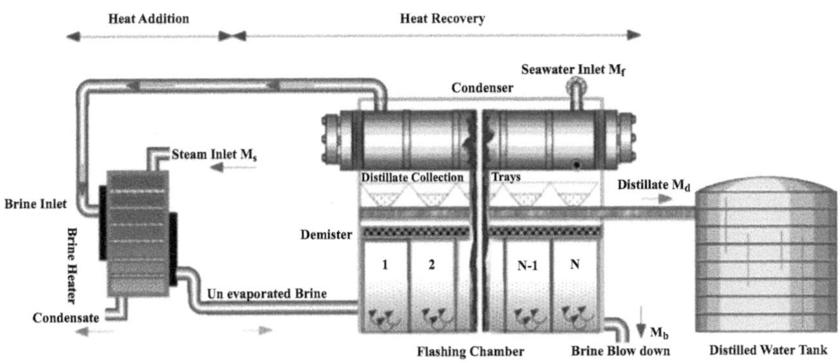

Figure 7.3 Illustration of the MSFD processes [22]. *MSFD*, Multistage flash distillation.

7.2.2 Multiple-effect distillation

The MED process is the oldest desalination method. Thermodynamically, MED is the most inherent method as compared to all others. The MED process takes place in a series of evaporators. The basic principle involved in MED is reducing the ambient pressure at different points. There is no need to provide or supply extra heat after the completion of the first effect as this process automatically allows the seawater feed to undergo multiple boilings.

As the seawater reaches the first effect, the temperature of the seawater is raised to boiling point after being preheated in the tubes. Thereafter, to carry out rapid evaporation, seawater is sprayed onto the surface of an evaporator. The dual-purpose power plant is used to supply steam externally, and the tubes are heated. Condensation of steam occurs on the opposite side of the tube and forms a steam condensate. Steam condensate is again utilized as boiler feed water after it is recycled to the power plant. The cyclic process of hybrid solar distillation is shown in Fig. 7.4.

For every plant, its economy balance is necessary for its smooth running. Similarly, for an MED plant, its economy is the key factor that depends directly on its number of effects. Basically, an MED is comprised

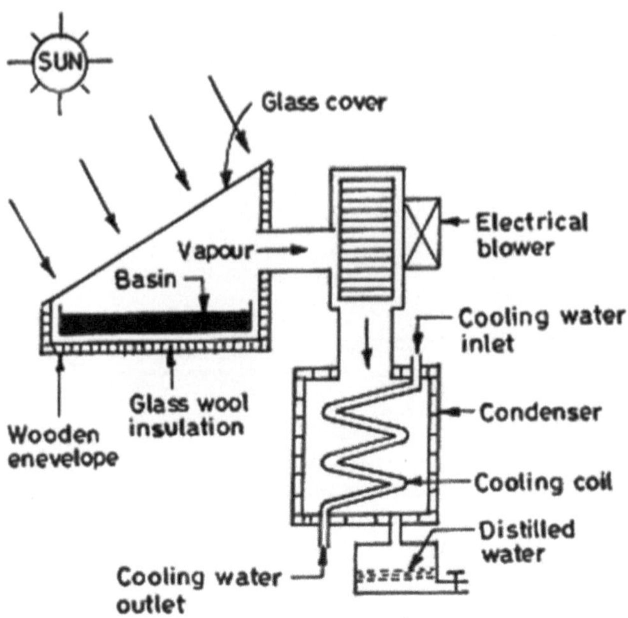

Figure 7.4 Distillation using solar energy [2].

of chained processes [8]. First, evaporation of some portion of seawater in the tubes takes place as the first effect. Then, the rest of the seawater is applied to the tube as it is fed into the second effect. Amid tubes are heated by the water vapor formed in the first effect. As a result, this vapor is condensed to the required product, which is freshwater. During the production of freshwater, vapor gives up its heat to evaporate the remaining seawater at the next effect. Repetition of the evaporation and condensation process takes place from one effect to another. This takes place successively at lower temperatures and pressures. This process continues and makes a chained process. This continues for about 4–21 effects [26].

The top brine temperature (TBT) is one of the most important factors for an MED plant. In an MED plant, it is necessary to reduce scale formation of seawater in the tubes. To maintain this very challenge, most of the plant is built to operate at a temperature of 70° TBT. However, to carry out the processes smoothly, an additional heat transfer area is required which is generally fulfilled by tubes. The performance ratio of an MED plant ranges from 10 to 18. From the context of thermodynamics and the heat transfer point of view, MED is better than MSFD as it needs less power and gives better performance.

A horizontal MED plant is more frequently found than other types of plant [25]. Moreover, these have been successfully operating for 30–40 years. Various types of tubes can be used in MED plants, such as vertical, horizontal, and submerged types.

7.2.3 Vapor compression distillation

In vapor compression distillation (VCD), compression of vapor plays an important role in providing heat to carry out the evaporation of seawater. The basic principle of VCD is reduction of the boiling point temperature by the reduction of pressure [27]. Due to this principle, VCD has an advantage over other methods. Mainly, two methods are used to carry out VCD—these are steam jet and mechanical compressor. These two methods are used to condense water vapor so that sufficient heat can be produced to carry out evaporation of seawater. Here, the electrically driven method is enabled by a mechanical compressor.

There are various types of configuration of VCD units. This allows easy exchange of heat to carry out evaporation of seawater. The steam jet type of VCD is also called a thermocompressor. In the thermocompressor, there is a venturi orifice at the steam jet which creates and extracts water vapor from the evaporator, thus creating a lower ambient pressure. Thereafter, the steam jet compresses the water vapor extracted by the venturi orifice. Then condensation of the mixture takes place on the wall of tube, which provides thermal energy to evaporate the seawater.

The other type of VCD is the low-temperature VCD which requires only power. Hence, it is a very simple, efficient, and reliable process. This method is applicable for mainly small-scale units of desalination. There are various application areas of VCD units such as resorts, drilling sites, and industries. It is very beneficial as it can be used where there is a lack of freshwater [28].

7.2.4 Reverse osmosis

The basic principle involved in this process is that osmotic pressure is to be overcome. To overcome osmotic pressure, external pressure is applied which is greater than that of the osmotic pressure on seawater. In an RO process, the flow of water reverses the direction of natural flow across the membrane; consequently, dissolved salt is left behind with an increased salt concentration. In this process, there is no need for phase separation or heating. Here, energy is required to pressurize the seawater feed to carry out the desalting process [29]. The key factors of an RO plant are its major components. There are four major components of RO plant, and these are feed water pretreatment, membrane separation, high-pressure pumping, and permeate posttreatment.

There are various undesirable components in seawater which can damage the membrane. Therefore there is a need to eliminate those constituents which is done by pretreatment [30]. Pretreatment of water includes various methods such as coagulation, chlorination, acid addition, multimedia filtration, micron cartridge filtration, and dechlorination. Pretreatment of the seawater feed depends on different factors like membrane type, feed water characteristics, membrane configuration, recovery ratio, and product water quality. Depending on these factors, the type of pretreatment is selected.

The key factor of an RO process is to reject the salt present in the seawater, for which the pretreated feedwater pressure is raised to the extent that it should be suitable for the RO membrane and water can easily pass through them. To raise the pressure, high-pressure stainless-steel pumps are used. The membrane should be such that it can bear the entire pressure drop across it. Generally, centrifugal pumps are used to carry out this operation. An illustration of a one-stage RO is shown in Fig. 7.5.

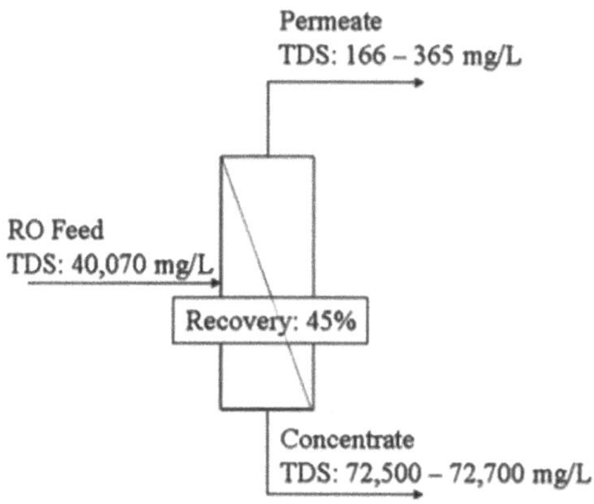

Figure 7.5 Schematic of a one-stage seawater RO [16]. *RO*, Reverse osmosis.

There are various membrane configurations, which spiral wound and hollow fine fiber (HFF) are most the economical and commercially successful. The shape of HFF is like a U-shaped fiber. In HFF, cellulose triacetate and polyamide are used as membrane materials [31].

Posttreatment is one of the most important components of RO process. In posttreatment mainly the pH is adjusted, lime is added, and dissolved gases like H_2S and CO_2 are removed.

Two changes and developments which were introduced in the previous decade have reduced the operating cost of RO plant. Those two developments were: the development of a membrane to operate efficiently for a longer time, and an energy recovery device [32]. These devices are used for the purpose of converting a pressure drop into rotating energy; hence, these devices are mechanical in nature.

7.2.5 Freezing

Desalination of seawater can also be done by a freezing process. It is a very simple process to obtain freshwater from seawater. During freezing, ice crystals are formed. Due to the formation of ice crystals, dissolved salts are removed. This can be achieved under a controlled situation. The mixture of seawater is generally cleaned and washed to remove salt in the water, just before the completion of freezing of whole water. Thus, freshwater is obtained by melting frozen ice. Hence, desalination by freezing includes various processes such as seawater cooling, then partial crystallization of ice, and thereafter ice is separated from seawater, before most importantly melting of ice to obtain freshwater, and finally the finishing processes of refrigeration and heat rejection [33].

Desalination by freezing has many advantages. Some of these include lower power consumption, less corrosion, and less scaling and precipitate formation. Along with these advantages, there are disadvantages also, like handling water and ice mixtures. It is very difficult to handle these two mixed together as they are mechanically difficult to process.

Although this process has many advantages, it has not been accepted commercially so far for the mass production of freshwater. Very few plants have been made till now, illustrating its unreliable nature. The famous plant till now was constructed in Saudi Arabia in 1985. This plant was an experimental solar-powered unit [34]. To govern the plant status, some processes were developed such as hydrate, indirect, eutectic, triple point, and secondary refrigerant processes.

7.2.6 Solar distillation

Solar desalination is one such process of desalination in which solar energy is the primary energy source to carry out the desalination of seawater. In this process, solar energy is used directly for desalinating seawater. Moreover, this process is similar to the process of the hydrological cycle. In the hydrological cycle, water vapor is produced by heating seawater with the sun's ray, and then, condensation of vapor takes place which ultimately gives condensate which is further collected as product water. Greenhouse solar desalination is one example of this type of desalination process [35].

This process was developed to increase the efficiency of the solar still, but it was found that it requires a large solar collection area of approximately 25 ha land/1000 m^3 of product water/day [28]. Not only space but also the high capital cost and vulnerability to weather-related damage are also disadvantages.

7.2.7 Potabilization

Potabilization is also a desalinating process, although it is an additive process to MSF. In other words, it is an add-on to MSF. When MSFD is completed, there is a small amount of impurities like dissolved salts and minerals, therefore the produced desalinated water is a little corrosive to the metals used in the water distribution system. To avoid these problems, potabilization is practiced [36].

There are two main typical methods carried out in the potabilization process. These are injection of carbon dioxide and hydrated lime [37], and carbonated water passed through limestone bed filters [38]. Basically, there are four processes involved in the potabilization process—carbonation, liming, chlorination, and aeration. The liming and carbonation processes signify remineralization of water with the addition of carbon dioxide and hydrated lime. The basic objective of liming and carbonation is to increase hardness, alkalinity, mineral content, and pH. There is a need to eliminate bacterial growth to disinfect the water. To avoid water contamination, a chlorination process is carried out [39]. This is done by injecting either chlorine gas or calcium hypochlorite. Last but not least, an aeration process is carried out to replace oxygen to improve the water taste.

7.3 Economics related to desalination

The economy of a plant depends primarily on production cost, location of plant, maintenance cost, energy consumption, etc. Due to changes in technologies and advancements in desalination, the cost of production is decreasing gradually but, on the other hand, due to the greater contamination of water because of population, pollution, etc., the cost of water treatment is also increasing due to the high demand for pure water.

The key factors in selecting a process between the two major processes for desalination of water (RO and MSFD) are technical and economic conditions. Some of the technical conditions which are taken into account are energy source, energy consumption, freshwater quality, space for plant, plant reliability, operational aspects, plant size, etc. Economic conditions that are taken into account are based on capital, labor, materials, chemicals, etc. [40].

For a desalination plant, a cogeneration scheme is necessary in conjunction with the generation of power for the best economy of the plant. The economics of a desalination plant are determined in the life cycle cost analysis. To evaluate the annual plant cost, O&M, that is operation and maintenance costs, are converted into annual costs. The cost of production of water is evaluated by dividing the sum of all costs by the total quantity of water produced. There are various parameters which affect the life cycle cost analysis such as plant life, direct capital cost, indirect capital cost, and capacity of production. This cost of production of freshwater is estimated in $\$/m^3$. For example, the world's largest RO plant has a water production cost of $\$0.53/m^3$.

7.4 Future expectance

Over the last two decades, there have been various improvements in the desalination of brackish and brine water. Many new technology and advancements have been introduced to increase the production of water through desalination. Economic conditions have also improved due to a reduction in water production costs. It has been accepted due to advancements in technology mainly in arid regions of the world [8]. To minimize the cost of production of water from desalination, further R&D has been taken into account. There is a need to emphasize R&D in technological advancements to improve the plants' economy. There are various parts and factors linked to desalination where R&D efforts impact and some of these are [41]: cogeneration system of desalinated water and power, energy utilization, mainly solar energy and nuclear energy, thermal distillation process at very high temperature, technical aspects of different methods of desalination, chemical therapy for seawater feed, economy of different desalination processes, best choice of construction and manufacturing materials, promotion of large-scale plants, scale-controlling system, cost-effective materials, introduction of hybrid systems like nanofiltration (NF)-RO, MSF-RO, etc., environmentally friendly desalination, and the percent separation of seawater and freshwater.

7.5 Solar still

Water is the most important component of our planet. It covers about 75% of the Earth, but of that amount only 1% can be used for domestic purposes, with some being contaminated by various factors like pollution, sewage disposal, etc. There is a need to obtain freshwater; however, most of the water present on the Earth is brackish and salty. Desalination is one of the measures to obtain freshwater from brackish water. To utilize desalination as an important measure, the solar still has been introduced in the developing world. The solar still is a device which is completely based on the principle of desalination. It mainly uses the concept of solar distillation. It is now being used worldwide, mainly in coastal areas where seawater is available in abundance. It is a simple, cost-effective, and easily maintained process [42]. The main disadvantage of the solar still is that it has lower productivity. Various researches and developments are on-going to enhance its efficiency.

For the development and modification of solar stills, various researches are being carried out on the basic design of the solar still to increase its productivity and make it more cost-effective [43]. The main idea for increasing the productivity of solar stills is increasing the heat transfer rate. To implement this idea, many researchers have used fins [44].

Srithar and Mani are famous scientists who have worked a lot on the development of solar stills. They dealt with the evaporation rate of industrial effluents. They developed a pilot plant for augmenting the evaporation rate. They also developed a pilot plant in two stages, one with spray network systems and another with an open fiber reinforced plastic flat plate collector. Then, they analyzed the performances separately to select the better one [45]. Sometimes, the performance was judged by means of sponges.

There are various components of the solar still like glass cover, container, basin liner, and trough. In a solar still, black paint is used to coat the inside surface of container; then, the collector is connected to a glass cover. Saline and brackish water fills the container for purification. The glass cover helps the radiation from the sun to be transmitted and later absorbed by the basin, which further heats the impure water. Thereafter, condensation of evaporated water takes place below the glass surface. Thus, it gathers in a trough which is located along its length. A simple basin solar still is shown in Fig. 7.6 [1].

Fig. 7.7 shows a schematic diagram of a solar still with a basin. This solar still consists of different parts for different applications. Various components of this solar still with a simple basin include the storage tank, valve, wooden box, hose, glass cover, still basin, collection tray, and measuring jar. The working of this solar still is very simple. The storage tank is the reservoir of water. It can have different capacities based on the requirements. Water from the tank reaches the still through two media—one is flexible hoses

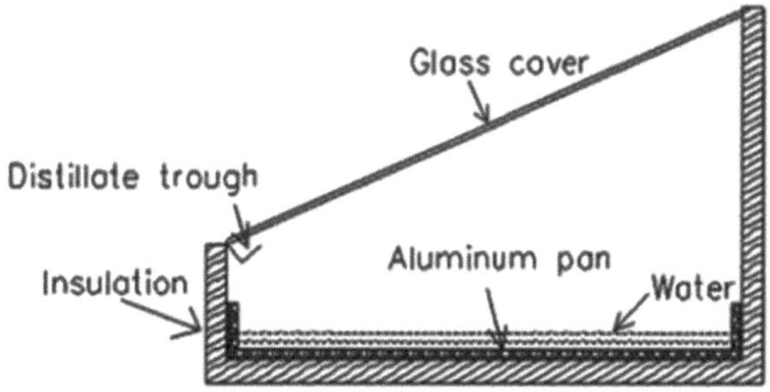

Figure 7.6 Single-basin solar still [46].

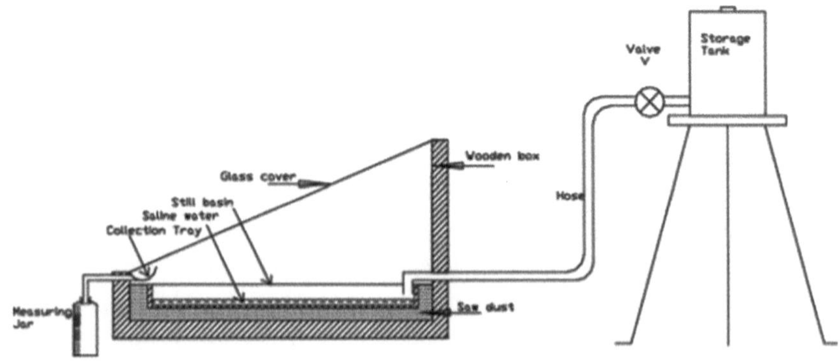

Figure 7.7 Solar still with a simple basin [44].

and the other is a valve. The role of the valve is to control the flow of water. The hoses are very flexible, and to maintain their flexibility they are made of polyvinyl chloride.

Here, the still basin is painted black and enclosed within a wooden box. The wooden box acts as a casing for the still. There is a need for insulation in the still, hence, sawdust is used. It is filled below the still basin. To collect the condensate after condensation of the evaporated water, a collection tray is used which is fixed to the wooden box, and thus freshwater is collected.

7.6 Types of solar still

There are basically two types of solar stills which are based on the effect, and these are single- and multieffect solar stills. These two types of solar stills are further subdivided into active and passive stills, which are categorized based on the source of heat provided to carry out the evaporation of water [1]. In one type, evaporation of water takes place directly but, in the other type, an external medium is required such as a heat exchanger or solar collector to carry out the evaporation of water. There are basically two types of solar still, as follows.

7.6.1 Single-effect solar still

The origin of the solar still development is the single-effect solar still. This is also called the original solar still. It is the simplest of all solar stills [47].

In this type of solar still, there is only one layer of glazing over the surface of the water. This characteristic of single-effect solar has proved to be one of its advantages. Due to the presence of a single layer of glazing, a large amount of heat loss takes place, thus reducing its efficiency. This heat loss takes place in the form of conduction. Thus, the efficiency of this type of solar still is about 30%–40% [1]. This type of solar still is also known as a single-slope solar still. Many experiments and studies have been carried out to improve its efficiency. The experimental setup for a single-effect solar still is comprised of various components like glass cover, measuring devices, basin liner, insulating materials, and distillate channels. Every component has different functions, and all work together to carry out an easy and proper desalination process. A simple single-effect solar still is shown in Fig. 7.7 with a full illustration. The working and function of different components of a single-effect solar still are as follows:

- Glass cover: This is the most important component of a solar still. It is set up at an angle to the horizontal. Generally, to strengthen the contact between the glass and other components of the solar still, silicon rubber is used. Another important additive is sealant, which acts as a frame to the still. This resists and compensates for any expansion and contraction between different materials.
- Basin liner: This is the base of the solar still. It has many properties such as high resistance to hot brackish water, high absorptivity, etc. Its main task is to absorb the radiation coming from the glass cover. It can be easily mended if damaged. Generally, asphalt is used as the basin liner.
- Measuring devices: In a solar still, there is a need to measure two parameters, one is temperature and the other one is wind speed. A digital anemometer is used to measure the wind speed. Thermocouples are generally used to measure temperature at different locations. Here thermocouples are connected to a digital thermometer. Usually, five thermocouples are used to measure the temperature at five locations, such as vapor, water, basin, and glass inside and out. Also, the intensity of solar radiation is measured using a heliometer.

Single-effect solar stills are further divided into two categories, active and passive solar stills, which are described next.

7.6.1.1 Active still

In a single-effect solar still, the active still deals with a source of heat which is external, such as industrial waste heat or solar collectors [1].

There are various types of active stills used in a single-effect solar still, and these are:

- Regenerative active solar still;
- Air-bubbled solar still;
- Waste heat recovery active solar still;
- Solar still with heat exchanger;
- Solar still integrated with solar concentrator;
- Solar still coupled with hybrid system;
- Solar still integrated with solar heaters.

Fig. 7.8 provides an illustration of an active still which is coupled to an evacuated tube collector.

7.6.1.2 Passive still

A passive still differs from an active still by the source of heat provided to evaporate the brackish water. In a passive still, internal heat from the still is taken in order to carry out evaporation of brackish or brine water [1]. There are different types of passive solar still for single-effect solar still, as follows:

- Basin solar still;
- Wick solar still;
- Weir-type still;
- Spherical still;
- Tubular still;

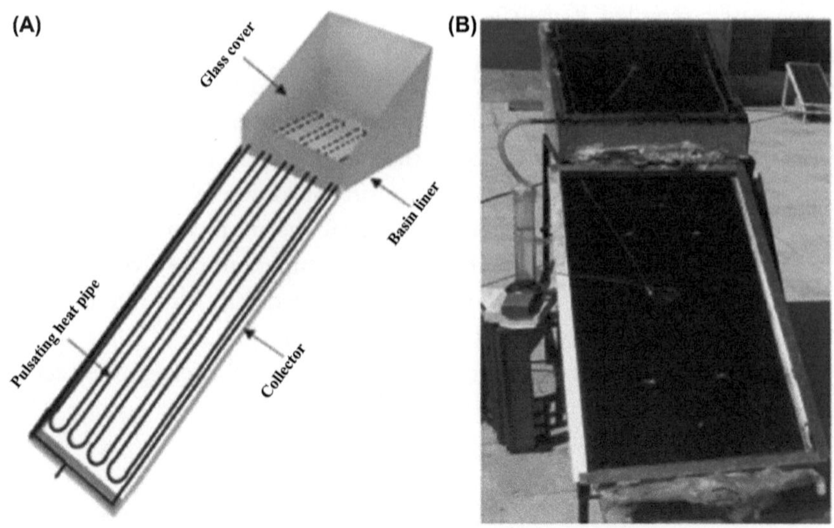

Figure 7.8 Active still coupled to an evacuated tube collector [48].

- Pyramidal and rectangular still;
- Diffusion still;
- Greenhouse combination solar still.

In Figs. 7.9 and 7.10, a passive still with different arrangements is shown. Fig. 7.9 shows a passive still arrangement which is coupled with an outside condenser, and Fig. 7.10 illustrates a passive solar still which is coupled with internal and external reflectors showing both the schematic and experimental setup.

7.6.2 Multieffect solar still

A multieffect solar still is another type of solar still which is quite different from a single-effect solar still. A multieffect solar still is more efficient than a single-effect solar still. Latent heat of condensation plays an important role in

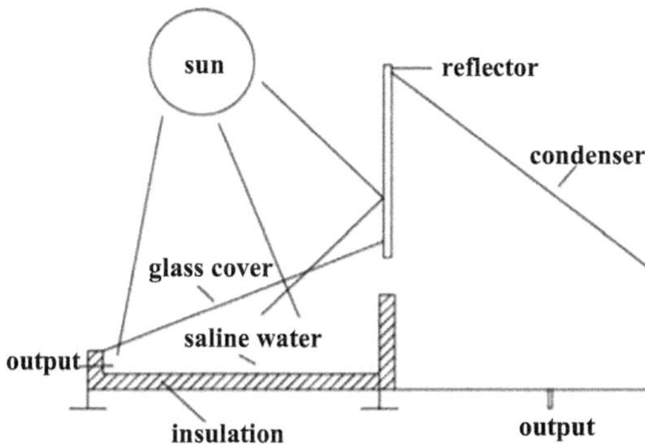

Figure 7.9 Passive still coupled with an outside condenser [49].

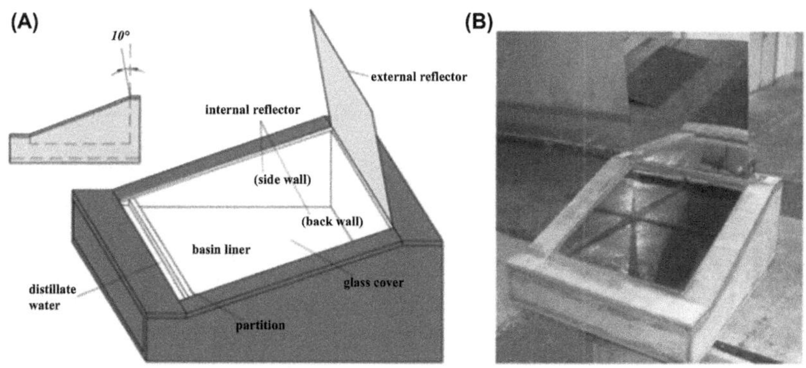

Figure 7.10 Passive still coupled with internal and external reflectors: (A) schematic diagram and (B) experimental setup [50].

the case of a multieffect solar still. In a multieffect solar still, recovery of latent heat of condensation takes place which is further recycled and thus increases the potential with a high rate of production [51]. This is also further classified as passive and active stills based on the design, as described next.

7.6.2.1 Active still

In an active multieffect solar still, the basic fundamental principle is the same as that of a single-effect one, with the only difference being in context of design. Different types of active multieffect solar stills are as follows:

- Multistage evacuated active solar still;
- Multibasin inverted absorber active still;
- Waste heat recovery active still;
- Solar still coupled with concentrating solar collectors;
- Multieffect condensation–evaporation water distillation system;
- Stills coupled with solar collectors like flat plate and tube collectors.

In Fig. 7.11 a multieffect active still with a single slope coupled with a solar collector in a thermosiphon and forced circulation mode is shown schematically. In Fig. 7.12, a condensation–evaporation active still system is shown.

7.6.2.2 Passive still

In the case of a multieffect solar still, there is only a difference in the context of design when a passive still is considered, otherwise the whole fundamental principle is completely the same. There are different types of passive multieffect solar still, as follows:

- Wick solar stills;
- Basin solar stills;
- Weir-type solar stills;
- Diffusion solar stills.

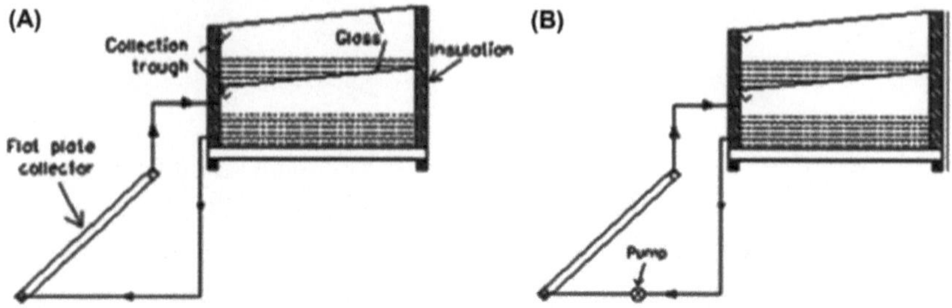

Figure 7.11 Double-effect single-slope active still: (A) coupled with a solar collector in thermosiphon mode and (B) coupled with a solar collector in forced circulation mode [52].

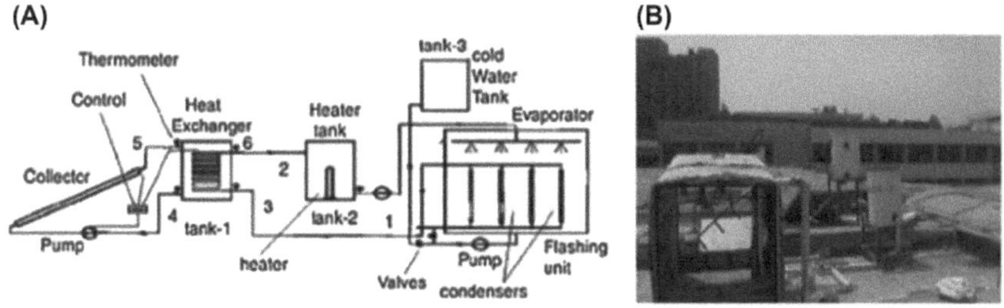

Figure 7.12 Condensation–evaporation active still system: (A) schematic diagram and (B) experimental setup [53].

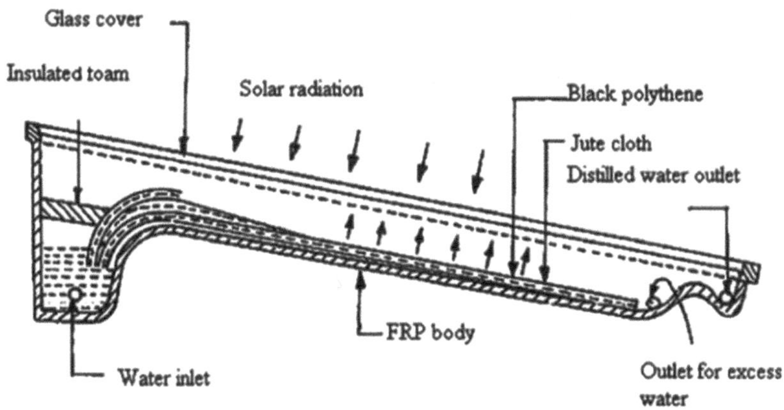

Figure 7.13 Wick-type passive solar still [45].

In Fig. 7.13, a passive still with wick-type solar still is depicted and in Fig. 7.14 a double-basin double-effect passive still is shown schematically.

7.7 Artificial neural network as a prediction method for the performance of desalination systems

Artificial intelligence is a rapidly advancing area of computer science. Since the invention of computers, we have observed an intriguing phenomenon. Tasks that are difficult for a human (such as heavy computations, searching, memorizing large volumes of data, and so on) are easily performed by computers, while tasks that humans are naturally able to do quickly (such as recognizing partially covered objects, intelligence, reasoning, creativity, invention, understanding speech, scientific research, and so on) are difficult for computers.

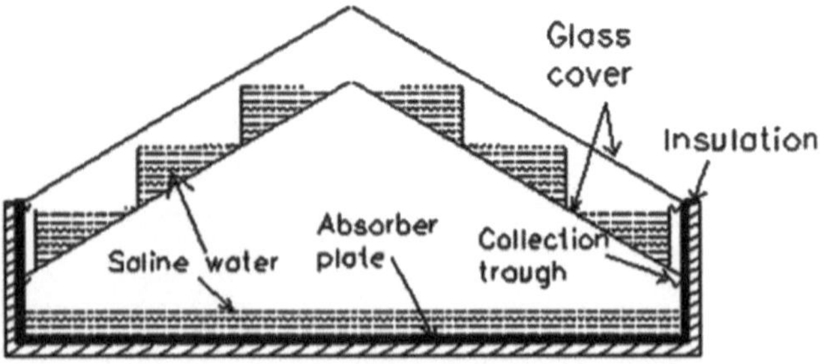

Figure 7.14 Double-effect passive still with double basin [46].

Artificial intelligence as a discipline was invented in the 1950s. Initially it failed because of a lack of backpropagation and an automated means of training. In the 1980s, it failed because of an inability to form its own internal representations, which was solved later by deep learning and the availability of more powerful computers.

A new nonlinear network architecture was developed after that second failure, and the tremendous increase in machines' computing power finally contributed to the phenomenal success of AI in the 1990s. Gradually, AI became capable of solving many industrial-grade problems such as image recognition, speech recognition, natural language processing, pattern recognition, prediction, classification, self-driving cars, robotic automation, and so on.

The tremendous success of AI has recently triggered all types of unwarranted speculation. You can find discussions about the robots of the near future matching and exceeding the intelligence of humans. However, currently, AI is a set of clever mathematical and processing methods that allow computers to learn from the data they process and apply this knowledge to solve many important tasks. Many things that belong to humans such as intelligence, emotion, creativity, feeling, reasoning, and so on, remain outside the possibility of AI.

Still, AI is rapidly changing. In recent years, computers have become so good at playing chess that they can reliably beat their human counterparts. That is not surprising, because their creators taught the programs centuries of accumulated human experience in chess. Now, machines compete against each other in the world computer chess championship. One of the best chess-playing programs, called Stockfish 8, won the world computer chess championship in 2016.

In 2017, Google developed a chess-playing program called AlphaZero, which defeated the Stockfish 8 program in the 2017 world computer chess championship. The amazing thing about this is that no one taught AlphaZero the chess strategies, as had been done during the development of other chess-playing programs. Instead, it used the latest machine learning principles to teach itself chess by playing against itself. It took the program 4 hours of learning chess strategies (while playing against itself) to beat Stockfish 8. Self-teaching is the new milestone achievement of artificial intelligence.

AI has many branches. This book is dedicated to one of them: neural networks. Neural networks enable computers to learn from observational data and make predictions based on that knowledge. Preparing data for neural network processing is typically the most difficult and time- consuming task you will encounter when working with neural networks. In addition to the enormous volume of data that could easily reach millions and even billions of records, the main difficulty is in preparing the data in the correct format for the task in question.

7.7.1 Network architecture

How is a network built? A network is built by including layers of neurons. The first layer on the left is the input layer, which contains the neurons that receive input from the outside. The last layer on the right is the output layer, which contains the neurons that carry the output of the network. One or more hidden layers are located between the input and output layers. Hidden-layer neurons are used for performing most of the calculations during the approximation of the function. Fig. 7.15 shows a network diagram.

Connections are drawn from the neurons in the previous layer to the neurons of the next layer. Each neuron in the previous layer is connected to all the neurons of the next layer. Each connection carries a weight, and each weight is numbered with two indexes. The first index is the receiving neuron number, and the second index is the sending neuron number. For example, the connection between the second neuron in the hidden layer (H_2) and the only neuron in the input layer (I_1) is assigned the weight W_{21}^1. The superscript 1 indicates the layer number of the sending neuron. Each layer is assigned a bias, which is similar to the weight assigned to a neuron but applied to a layer level. Biases make the linear part of each neuron output calculation more flexible in matching the approximated function topology.

When the network processing starts, the initial values for weights and biases are usually randomly set. Typically, to determine the number of neurons in a hidden layer, you double the number of neurons in the

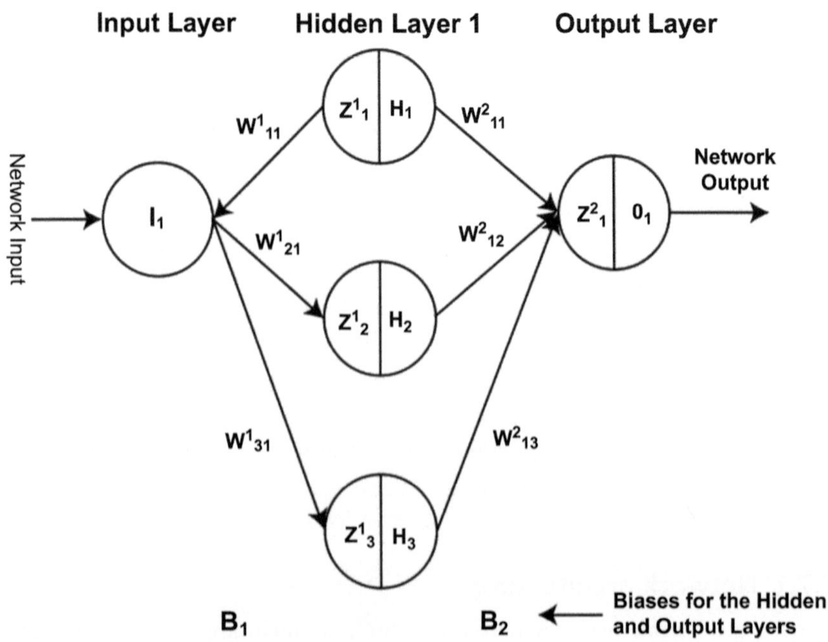

Figure 7.15 Neural network architecture.

input layer and add the number of neurons in the output layer. In our case, it is ($1 \times 2 + 1 = 3$), or three neurons. The number of hidden layers to be used in the network depends on the complexity of the function to be approximated. Typically, one hidden layer is sufficient for a smooth continuous function, and more hidden layers are needed for more complex function topology. In practice, the number of layers and neurons in the hidden layers that leads to the best approximation results is determined experimentally.

The network processing consists of two passes: a forward pass and a backward pass. In the forward pass, the calculation moves from left to right. For each neuron, the network gets the input to the neuron and calculates the output from the neuron.

7.7.2 Application of artificial neural networks in desalination systems

Contemporary experience indicates that artificial neural networks (ANNs) may be specifically appropriate to offer tools to assist desalination plant operators in day-to-day operations. Forecasting the productivity of desalination units is a critical issue to reduce the burden of laboratory equipment on these units. Artificial intelligence-based methods have been reported as robust tools to figure out the correlation between the process factors and responses of different engineering systems [54–58]. This chapter suggests that ANNs are particularly appropriate as the basis for the development of tools to aid in the various phases of operating a desalination plant. In solar energy applications, different types of ANNs succeeded to predict and optimize the performance of solar collectors, solar cells, solar stills, and others [59]. It is noted that there has been significant progress in applying neural networks to the problems of planning and operations in electric power systems. Realizing the close similarities between power system and desalination plant operational requirements, this chapter goes on to suggest potential applications in desalination plants. The features of each problem area are first defined from a desalination plant operator's perspective. This is followed by a discussion of approaches available in a similar power systems area. The treatment is limited to areas such as alarm processing, fault detection, control applications, operational optimization, load forecasting, and security assessment [60]. The main advantages of using ANNs include: they can handle large amount of data sets; they have the ability to implicitly detect complex nonlinear relationships between dependent and independent variables; has and they have the ability to detect all possible interactions between predictor variables.

Zarei and Behyad [61] employed an ANN to examine the effective parameters of a greenhouse on freshwater production, such as width, length, the height of the front evaporator, and roof transparency. The method showed good agreement with the experimental data. Tayyebi and Alishiri [62] used a nonlinear inverse model control strategy based on a neural network for predicting the performance of an MSF desalination plant. The designed neural networks consisted of three layers identified from the input–output data and trained with a descent gradient algorithm. The set point tracking performance of the proposed method was studied when a disturbance was present in the MSF system. Three controllers are designed for controlling the TBT, the level of last stage and salinity. These results show that a neural network inverse model control strategy is robust and highly promising to be implemented in such nonlinear systems. An ANN model based on the experimental data was developed to study the performance of a vacuum membrane distillation (VMD) desalination process under different operating parameters such as the feed inlet temperature, the vacuum pressure, the feed flow rate, and the feed salt concentration [63]. The proposed model was found to be capable of predicting accurately the unseen data of the VMD desalination process. The correlation coefficient of the overall agreement between the ANN predictions and experimental data was found to be more than 0.994. Aish et al. [64] used an ANN technique to forecast an RO desalination plant's performance in the Gaza Strip by predicting the next week's values of total dissolved solids (TDS) and permeate of the product water. Multilayer perceptron (MLP) and radial basis function (RBF) neural networks were trained and developed with reference to feed water parameters including: pressure, pH, and conductivity to predict permeate of the next week values. MLP and RBF neural networks were used for predicting the next week's TDS concentrations. Both networks were trained and developed with reference to product water quality variables, including water temperature, pH, conductivity, and pressure. The prediction results showed that both types of neural networks were highly satisfactory for predicting the TDS level in the product water quality and satisfactory for predicting permeate amount.

Santos et al. [65] modeled solar still productivity based on local weather data, such as total solar radiation, average ambient temperature, average cloud cover, average wind velocity, and average wind direction, using ANNs. Essa et al. [66] proposed an enhanced version of an ANN which incorporated with the Harris Hawks optimizer to predict the water yield of an active solar still. The proposed optimizer was used to obtain the optimal structure and parameters of the neural network. However, conventional ANN s suffer from trapping in local optima problem during the learning process. To overcome this problem random vector functional link (RVFL) is proposed as it determines the output weights without involving a time-consuming learning process. In addition, RVFL has a direct link between the input and output nodes, which has a significant effect on the performance of the network. This helps to avoid the overfitting problem that occurs in other ANNs. With these advantages, the traditional RVFL networks still face some limitations such as different configurations can be used, and this can affect the quality of the final prediction of water production. So, Ensemble Random Vector Functional Link Networks (EnsRVFL) [67], which depend on the RVFL as a basis model, have been used to avoid these limitations which uses RVFL as a base model. An RVFL network integrated with an artificial ecosystem-based optimization (AEO) algorithm is proposed to predict the performance of the seawater greenhouse (SWGH) system [68]. Power consumption and water productivity of the

SWGH are predicted using the proposed RVFL-AEO model. The statistical analyses using different statistical criteria such as root mean square error, mean absolute error, mean relative error, efficiency coefficient, coefficient of determination, overall index, and coefficient of residual mass are also carried out to examine the efficiency of the proposed neural network. The statistical tools obtained a perfect match between the experimental and the proposed model results. The performance of the RVFL-AEO model was compared with that of the conventional RVFL model. RVFL-AEO showed a better performance compared with RVFL; which indicates the role of AEO in obtaining the optimal RVFL parameters that enhance the accuracy of the model.

7.8 Conclusions

Water is a necessity for human life. The rapidly increasing pollution and population has resulted in contaminated water. The availability of freshwater is limited on Earth; therefore there is a need to obtain more freshwater for the survival of humans and for our day-to-day activities. Desalination has been proved to be the best measure to obtain freshwater, as salty water covers almost 70% of the Earth's surface. To make best use of desalination, a solar still can be used as an efficient device to obtain increasing amounts of freshwater. Various researches are on-going in the solar still field. The following points which can be concluded from this review:

- Desalination is the most cost-effective way to enhance freshwater availability on Earth. The demand of freshwater is rapidly increasing daily.
- The demister plays a prominent role in MSFD as it is used to overcome the brine into the vapor system so as to maintain undisturbed purity of the water.
- Stainless steel and duplex steel have been accepted as the principal materials to make accessories for MSFD plants. Moreover, stainless steel reduced the weight and cost of freshwater production.
- For the construction of the ejector required in MSFD, Incoloy 825 Nickel was used as the chief material due its high pitting resistance equivalent number.
- The gained output ratio is higher for MED than MSFD (it is about 15 for MED and 10 for MSFD).

- RO is the most effective and efficient method for desalination. Pretreatment is the main issue to be solved smoothly in an RO plant, and to overcome this problem, new technology has arrived, called named NF membrane treatment. Moreover, NF technology has helped in increasing the production rate by 40% and avoids membrane fouling.
- The technical and economic conditions are the key factors in deciding which process should be used for desalination: RO or MSFD. These two processes are the main desalination processes used around the world.
- Life cycle cost analysis plays a vital role in deciding the economics for desalination plants.
- The solar still is the simplest device to obtain freshwater from abundant salty water. It works on the basic principle of solar distillation.
- The productivity of a solar still is not high, so there is a need to increase the heat transfer rate to increase the rate of production.
- Previously, single-plate solar stills were introduced in the market but later, after various researches and changes, the multieffect solar still arrived in the market with a greater efficiency and rate of production.
- The solar still is mainly categorized as active and passive stills. This classification is based on the source of heat provided to carry out the evaporation process.
- An active still takes heat from external sources such industrial waste or by using solar collectors, while a passive still takes heat internally from the still to carry out the evaporation of water.
- The configuration of a solar still is very simple. It is a very cost-effective device to obtain freshwater from brackish and brine water. It has been proved to be very beneficial for arid and semiarid regions where there is a greater shortage of freshwater as compared to other regions of the world. Various researches are on-going to increase its efficiency and introduce it as a large-scale production device.

References

[1] P.V. Kumar, et al., Solar stills system design: a review, Renewable and sustainable energy reviews 51 (2015) 153–181.
[2] G. Tiwari, H. Singh, R. Tripathi, Present status of solar distillation, Solar energy 75 (5) (2003) 367–373.
[3] F.R. Rijsberman, Water scarcity: fact or fiction? Agricultural Water Management 80 (1) (2006) 5–22.
[4] J.S. Wallace, Increasing agricultural water use efficiency to meet future food production, Agriculture, ecosystems & environment 82 (1) (2000) 105–119.

[5] C. Pahl-Wostl, Transitions towards adaptive management of water facing climate and global change, Water Resources Management 21 (1) (2007) 49–62.
[6] M. Kummu, et al., Is physical water scarcity a new phenomenon? Global assessment of water shortage over the last two millennia, Environmental Research Letters 5 (3) (2010) 034006.
[7] J.R. McCutcheon, R.L. McGinnis, M. Elimelech, A novel ammonia—carbon dioxide forward (direct) osmosis desalination process, Desalination 174 (1) (2005) 1–11.
[8] A.D. Khawaji, I.K. Kutubkhanah, J.-M. Wie, Advances in seawater desalination technologies, Desalination 221 (1–3) (2008) 47–69.
[9] H.M. Qiblawey, F. Banat, Solar thermal desalination technologies, Desalination 220 (1) (2008) 633–644.
[10] A. Abdullah, et al., Performance evaluation of a humidification–dehumidification unit integrated with wick solar stills under different operating conditions, Desalination 441 (2018) 52–61.
[11] S.W. Sharshir, et al., Performance enhancement of wick solar still using rejected water from humidification-dehumidification unit and film cooling, Applied Thermal Engineering 108 (2016) 1268–1278.
[12] F.A. Essa, et al., On the different packing materials of humidification–dehumidification thermal desalination techniques – a review, Journal of Cleaner Production (2020) 123468.
[13] A. Abdullah, et al., An augmented productivity of solar distillers integrated to HDH unit: Experimental implementation, Applied Thermal Engineering 167 (2020) 114723.
[14] S. Lattemann, T. Höpner, Environmental impact and impact assessment of seawater desalination, Desalination 220 (1) (2008) 1–15.
[15] K.J. Templer, C. Tay, N.A. Chandrasekar, Motivational cultural intelligence, realistic job preview, realistic living conditions preview, and cross-cultural adjustment, Group & Organization Management 31 (1) (2006) 154–173.
[16] L.F. Greenlee, et al., Reverse osmosis desalination: water sources, technology, and today's challenges, Water Research 43 (9) (2009) 2317–2348.
[17] K. Saito, et al., Power generation with salinity gradient by pressure retarded osmosis using concentrated brine from SWRO system and treated sewage as pure water, Desalination and Water Treatment 41 (1–3) (2012) 114–121.
[18] B.A. Kamaluddin, S. Khan, B.M. Ahmed, Selection of optimally matched cogeneration plants, Desalination 93 (1–3) (1993) 311–321.
[19] S. Consonni, G. Lozza, E. Macchi, Optimization of cogeneration systems operation. Part A: Prime movers modelization, in: Proc. of the 1989 ASME Cogen-Turbo Symposium, 1989.
[20] C. Sommariva, V. Syambabu, Increase in water production in UAE, Desalination 138 (1–3) (2001) 173–179.
[21] M. Finan, A. Harris, S. Smith, The field assessment of a high temperature scale control additive and its effect on plant corrosion, Desalination 20 (1–3) (1977) 193–201.
[22] H. Baig, M.A. Antar, S.M. Zubair, Performance evaluation of a once-through multistage flash distillation system: impact of brine heater fouling, Energy Conversion and Management 52 (2) (2011) 1414–1425.
[23] T. Younos, Environmental issues of desalination, Journal of Contemporary Water Research and Education 132 (1) (2005) 3.
[24] A. Al Mudaiheem, H. Miyamura, Construction and commissioning of al Jobail phase II desalination plant, Desalination 55 (1985) 1–11.
[25] T. Pankratz, Water desalination report – 31 July 2017. Water Desalination Report, Media Analytics, 2017.

[26] T. Michels, Recent achievements of low temperature multiple effect desalination in the western areas of Abu Dhabi. UAE, Desalination 93 (1–3) (1993) 111–118.
[27] N.R. Council, Committee on advancing desalination technology, Desalination: a National Perspective (2008).
[28] O. Buros, The ABCs of desalting, International Desalination Association Topsfield, MA, 2000.
[29] M.B. Baig, A.A. Al Kutbi, Design features of a 20 migd SWRO desalination plant, Al Jubail, Saudi Arabia, Desalination 118 (1–3) (1998) 5–12.
[30] A.H.H. Al-Sheikh, Seawater reverse osmosis pretreatment with an emphasis on the Jeddah Plant operation experience, Desalination 110 (1–2) (1997) 183–192.
[31] G.M. Whitesides, Whitesides' group: writing a paper, Advanced Materials 16 (15) (2004) 1375–1377.
[32] S. Avlonitis, K. Kouroumbas, N. Vlachakis, Energy consumption and membrane replacement cost for seawater RO desalination plants, Desalination 157 (1) (2003) 151–158.
[33] O. Buros, Conjunctive use of desalination and wastewater reclamation in water resource planning, Desalination 19 (1–3) (1976) 587–594.
[34] D. Hanahan, R.A. Weinberg, Hallmarks of cancer: the next generation, Cell 144 (5) (2011) 646–674.
[35] F. Chang, et al., Bigtable: A distributed storage system for structured data, ACM Transactions on Computer Systems (TOCS) 26 (2) (2008) 1–26.
[36] A. Withers, Options for recarbonation, remineralisation and disinfection for desalination plants, Desalination 179 (1–3) (2005) 11–24.
[37] A.D. Khawaji, J.-M. Wie, Potabilization of desalinated water at Madinat Yanbu Al-Sinaiyah, Desalination 98 (1–3) (1994) 135–146.
[38] H. Al-Rqobah, A recarbonation process for treatment of distilled water produced by MSF plants in Kuwait, Desalination 73 (1989) 295–312.
[39] Y. Ayyash, et al., Performance of reverse osmosis membrane in Jeddah Phase I plant, Desalination 96 (1–3) (1994) 215–224.
[40] N.M. Wade, Technical and economic evaluation of distillation and reverse osmosis desalination processes, Desalination 93 (1–3) (1993) 343–363.
[41] X. Yin, Z. Yang, Development of a coupled reservoir operation and water diversion model: balancing human and environmental flow requirements, Ecological modelling 222 (2) (2011) 224–231.
[42] A. Nafey, et al., Solar still productivity enhancement, Energy Conversion and Management 42 (11) (2001) 1401–1408.
[43] I. Al-Hayeka, O.O. Badran, The effect of using different designs of solar stills on water distillation, Desalination 169 (2) (2004) 121–127.
[44] V. Velmurugan, et al., Single basin solar still with fin for enhancing productivity, Energy Conversion and Management 49 (10) (2008) 2602–2608.
[45] K. Srithar, A. Mani, Open fibre reinforced plastic (FRP) flat plate collector (FPC) and spray network systems for augmenting the evaporation rate of tannery effluent (soak liquor), Solar Energy 81 (12) (2007) 1492–1500.
[46] T. Rajaseenivasan, et al., A review of different methods to enhance the productivity of the multi-effect solar still, Renewable and Sustainable Energy Reviews 17 (2013) 248–259.
[47] O.O. Badran, Experimental study of the enhancement parameters on a single slope solar still productivity, Desalination 209 (1–3) (2007) 136–143.
[48] H.K.S. Abad, et al., A novel integrated solar desalination system with a pulsating heat pipe, Desalination 311 (2013) 206–210.
[49] A. El-Bahi, D. Inan, A solar still with minimum inclination, coupled to an outside condenser, Desalination 123 (1) (1999) 79–83.

[50] H. Tanaka, Experimental study of a basin type solar still with internal and external reflectors in winter, Desalination 249 (1) (2009) 130–134.
[51] H. Tanaka, T. Nosoko, T. Nagata, Experimental study of basin-type, multiple-effect, diffusion-coupled solar still, Desalination 150 (2) (2002) 131–144.
[52] Y. Yadav, Transient analysis of double-basin solar still integrated with collector, Desalination 71 (2) (1989) 151–164.
[53] A.M.A. Dayem, Experimental and numerical performance of a multi-effect condensation–evaporation solar water distillation system, Energy 31 (14) (2006) 2710–2727.
[54] T.A. Shehabeldeen, et al., A novel method for predicting tensile strength of friction stir welded AA6061 aluminium alloy joints based on hybrid random vector functional link and henry gas solubility optimization, IEEE Access 8 (2020) 79896–79907.
[55] M.A. Elaziz, A.H. Elsheikh, S.W. Sharshir, Improved prediction of oscillatory heat transfer coefficient for a thermoacoustic heat exchanger using modified adaptive neuro-fuzzy inference system, International Journal of Refrigeration 102 (2019) 47–54.
[56] A.H. Elsheikh, et al., An artificial neural network based approach for prediction the thermal conductivity of nanofluids, SN Applied Sciences 2 (2) (2020) 235.
[57] T.A. Shehabeldeen, et al., Modeling of friction stir welding process using adaptive neuro-fuzzy inference system integrated with Harris Hawks optimizer, Journal of Materials Research and Technology 8 (6) (2019) 5882–5892.
[58] H.A. Babikir, et al., Noise prediction of axial piston pump based on different valve materials using a modified artificial neural network model, Alexandria Engineering Journal 58 (3) (2019) 1077–1087.
[59] A.H. Elsheikh, et al., Modeling of solar energy systems using artificial neural network: A comprehensive review, Solar Energy 180 (2019) 622–639.
[60] M. El-Hawary, Artificial neural networks and possible applications to desalination, Desalination 92 (1–3) (1993) 125–147.
[61] T. Zarei, R. Behyad, Predicting the water production of a solar seawater greenhouse desalination unit using multi-layer perceptron model, Solar Energy 177 (2019) 595–603.
[62] S. Tayyebi, M. Alishiri, The control of MSF desalination plants based on inverse model control by neural network, Desalination 333 (1) (2014) 92–100.
[63] W. Cao, et al., Modeling and simulation of VMD desalination process by ANN, Computers & Chemical Engineering 84 (2016) 96–103.
[64] A.M. Aish, H.A. Zaqoot, S.M. Abdeljawad, Artificial neural network approach for predicting reverse osmosis desalination plants performance in the Gaza Strip, Desalination 367 (2015) 240–247.
[65] N.I. Santos, et al., Modeling solar still production using local weather data and artificial neural networks, Renewable Energy 40 (1) (2012) 71–79.
[66] F.A. Essa, M. Abd Elaziz, A.H. Elsheikh, et al., An enhanced productivity prediction model of active solar still using artificial neural network and Harris Hawks optimizer, Applied Thermal Engineering 170 (2020) 115020.
[67] M. Alhamdoosh, D. Wang, Fast decorrelated neural network ensembles with random weights, Information Sciences 264 (2014) 104–117.
[68] F.A. Essa, M. Abd Elaziz, A.H. Elsheikh, et al., Prediction of power consumption and water productivity of seawater greenhouse system using random vector functional link network integrated with artificial ecosystem-based optimization, Process Safety and Environmental Protection 144 (2020) 322–329.

CHAPTER EIGHT

Artificial neural networks for engineering applications: a review

Mohammad Shehab[1,5], Laith Abualigah[2,4,5,6], Mahmoud Omari[6], Mohd Khaled Yousef Shambour[3], Mohammad Alshinwan[6], Hayfa Y. Abuaddous[6] and Ahmad M. Khasawneh[6]

[1]Information Technology, The World Islamic Sciences and Education University, Amman, Jordan
[2]Hourani Center for Applied Scientific Research, Al-Ahliyya Amman University, Amman, Jordan
[3]Faculty of Information Technology, Middle East University, Amman, Jordan
[4]School of Computer Sciences, Universiti Sains Malaysia, Pulau Pinang, Malaysia
[5]Faculty of Computer Sciences and Informatics, Amman Arab University, Amman, Jordan
[6]Department of Scientific Information and Services, Umm Al-Qura University, Mecca, Saudi Arabia

Contents

8.1 Introduction

The use of artificial neural networks (ANNs) has evolved significantly in various fields because it is considered to be a biologically inspired analytical technique [1−3]. In other words, AANs are based on the computer mechanism of the human brain, which is a totally different method

Artificial Neural Networks for Renewable Energy Systems and Real-World Applications.
DOI: https://doi.org/10.1016/B978-0-12-820793-2.00003-3

from a traditional computing machine [4,5]. For instance, an ANN dealing with segmental memories units and emulation approaches uses modern neuroscience (i.e., as neurons), such as motor control, perception, and pattern recognition [6–8]. Thus, there is no need for previous models, or previous assumptions for the properties of data [9–11]. Also, several optimization algorithms have been proposed recently which can be used to solve the various problems of engineering applications [12–14].

Fig. 8.1A shows the biological neuron which is responsible for transmitting the electrical signal from the sensory, audio, and visual receptors through the axon to the other connected neurons (dendrites). Meanwhile, Fig. 8.1B illustrates a simple method that is inspired from a biological neuron, in which it can be noted that there are n inputs related to the center (neuron). Two operations have been used to process the inputs; (1) teaching, also known as training, which provides the neurons with feedback to take the necessary action and (2) the using operation which is employed to detect the input pattern. The output is based on the number of inputs, as well as the two operations inside the neuron.

Moreover, backpropagation (BP)-based ANNs are utilized to predict the events before they happen from the environment [15–17]. For instance, it can predict a time delay, possible solutions, and the potential risks. Thus, it helps to achieve the optimal solution with low cost and minimum time. Fig. 8.2 shows a typical three-layer BP neural network model. It can be seen that the main function is converting the hidden layer nonlinear function based on the inputs and expected values to produce the actual outputs. This model has proved its efficiency in various domains, and the following paragraphs describe some of the usages of BP-ANN.

A feedforward BP-ANN was used to predict the academic student performance at the University of Al-Azhar in Gaza [18]. The authors

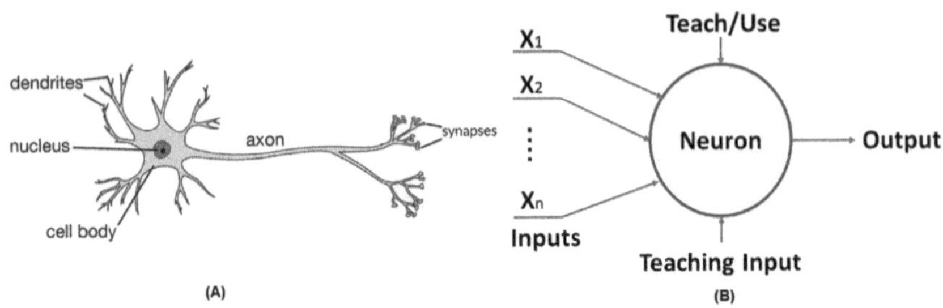

Figure 8.1 (A) A biological neuron. (B) A simple neuron.

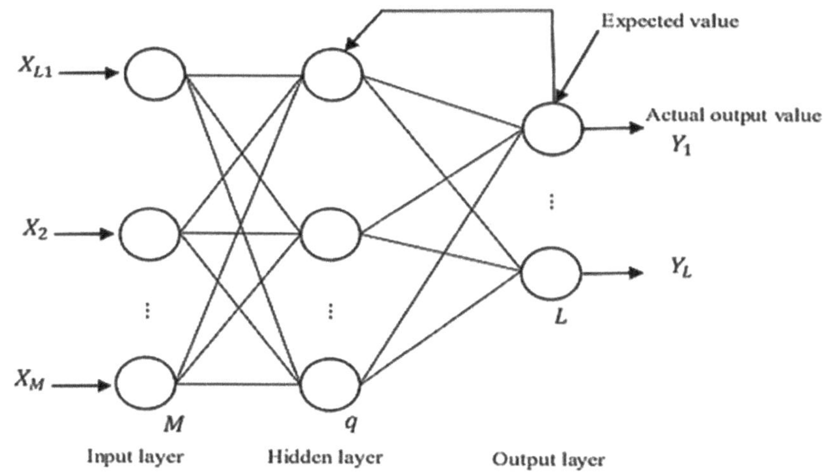

Figure 8.2 Typical BP neural network model.

defined 10 factors that affected student performance and used them as input for the system. The collected dataset contained 150 students' records, divided into three divisions. The first division contained 60% of the dataset, prepared for training phase; whereas the second and third divisions contained 30% and 10% of the dataset, respectively, used for testing and cross-validation phases. The results showed the ability of BP-ANNs to predict student performance with an accuracy of 84.6%.

A feedforward BP-ANN was applied to predict two environmental systems, air pollution and flood, in a certain hydrographic basin [19]. Historical data series of air pollutants (2005–07) and river flow (2005–08) were used for training the NN to predict the future values of a time series. Different architectures of ANNs were tested in an air pollution system to observe the best architecture with the lowest mean square error value. Moreover, results of the flood prediction system showed the advantages of ANNs over a classical mathematical model that required more processing time and efforts from hydrology experts.

The Kansei Engineering method was applied to formulate the worker capacity in a food production system [20]. The authors defined six factors for assessment process, including worker mood, humidity, workplace temperature, heart rate, noise level, and light intensity. These factors were measured daily, at the start and end of working times for the workers, and were prepared to be used as input for the BP-ANN to predict the worker capacity status.

Another worker evaluation model was introduced for a food production system using Kansei Engineering and ANN. The introduced model was comprised of worker capacity assessment and submodels of worker performance evaluation. The BP-ANN model was employed to predict the mood disturbance score of workers using the same factors as discussed in Refs. [21–25].

Based on the above discussion, some advantages of NNA can be summarized as follows:

- ANNs have the capability to learn and model nonlinear and complicated relationships [26].
- ANNs can generalize [27].
- ANNs have the potential for high fault tolerance [28].
- ANNs have the ability of parallel processing [29]

8.2 Application of artificial neural networks in engineering fields

8.2.1 Chemical engineering

Chemical engineering is a promising research area in engineering and computer science [30–33]. Mazrou Hakim proposed a new method to improve the safety of nuclear research reactors based on ANNs [34,35]. The neural network is used to predict two parameters in safety evaluation. The training process of the neural algorithm exceeded different levels, such as various transfer functions, different concealing neurons, and the levels of size and failure of generalization corpus. The result showed that a significant performance in the network generalization level and computational time during the training time.

In Ref. [36], the authors utilized a neural network to construct a chemical oxygen model for dairy manufacturing effluent after electrocoagulation. The impacts of operating parameters were assessed, including the electrical current weight, primary effluent pH, electrolysis rate, and gap between electrodes. Various parameters were ranked by their relevance for prediction of the chemical oxygen requirement after processing by electrocoagulation. Operating settings may be set to obtain a more significant decrease in the chemical oxygen needed and to support industrialization of the treatment method.

In the processes of the chemical reactor, it is challenging to identify the nonlinear behavior of chemical treatment. Thus, in Ref. [37], a hybrid neural network algorithm with a rate function called HNNRF with mechanistic equations has been proposed. The proposed method modeling capacity across a broad operating area was assessed by applying a simulated polymerization reactor method. The simulation result showed acceptable results in term of noisy measurements, and the classified model was utilized to define the optimal method or the operating requirements of the reactor system.

Li et al. introduced a continuous reactor for three-level catalytic slurry hydrogenation [38]. The control of the enhanced continuous reactor created challenges due to complicated three-level reaction kinetics, inherent process nonlinearities, and severe temperature restriction. This work proposed a nonlinear method for predictive control called NMPC, which was based on neural networks. The simulation results demonstrated that the proposed algorithm showed an outperformance. Other methods can be found in Refs. [39,40].

8.2.2 Civil engineering

Singh et al. [41] reviewed the different ANN architectures and their learning processes related to civil engineering. ANNs were employed in a civil engineering scenario with accurate results in a nonlinear relation between unknown and known aspects of the problems.

Recently, an alkali-activated substance used as a new construction material to reduce carbon and waste recycling. Different studies of the alkali-activated material showed that alkali-activated slag mortar (AASM) has low thermic, robust mechanic features, and a strong perspective for in situ utilization. Consequently, modeling for assessing the engineering features of AASM is essential. Therefore, Chen et al. [42] utilized an ANN to predict the engineering characteristics of concrete. The simulation results showed that the NNA prediction results outperformed the other methods described in the literature.

8.2.3 Computer engineering

Kaur et al. [43] discovered that ANNs have a learning machine capability and provide the ability to identify patterns. This system includes interconnected "neurons" used by feeding information within the network to determine values from inputs. In this paper, the authors present theoretical and practical studies of ANNs with the aim of understanding and evaluating the storage capacity of any network.

In Ref. [44], the authors focus on the design of a simple graphical user interface in MATLAB® using a neural network algorithm to predict the output for a set of inputs based on the learning example given. By taking speed, feed, and cut depth as input parameters from a CNC milling machine, the developed method was used to predict the material removal rate and tool wear. This method was also used to estimate an inboard fixed trailing edge rib volume and stress for various thickness parameter combinations. The expected outcome was then compared to the ANSYS results for error percentage estimation. A higher learning rate quality is used to estimate large amounts of data.

In Ref. [45], a study was conducted based on two fundamental objectives. It aimed to implement a smart software system that optimizes and enhances the self-learning processes of computer engineering students. Moreover, it focused on improving the self-learning of students in computer engineering during the course. As a general rule, the software system implemented here measured the intelligence levels of students according to the Multiple Intelligence Theory. It facilitated their learning by correctly selected materials supplied through the software interface. The system's evaluation method is based on a hybrid approach to artificial intelligence generated by an optimization algorithm called the vortex optimization algorithm (VOA) and ANN. The model can be used mainly for technical courses taught in computer engineering university departments and makes complex subjects easier to teach. To get an indication of the system's performance, students were tested, and positive results were obtained on maximizing and enhancing self-learning.

8.2.4 Power and energy engineering

Power and energy engineering is a promising research area in engineering and computer science [46,47]. In Ref. [48], a laboratory-based research study was performed at various levels of pollution on the system under ac voltage. To predict flashover from the analysis of leakage current, a new model based on the ANN was developed. The input parameters for the ANN included the mean (Imean), maximum (Imax), and standard deviation (I) of the derived leakage current along with the input voltage (V) and relative humidity (RH). The target observed has been utilized to assess the neural network model's performance. The simulated and actual (measured) results comparison showed that the ESDD prediction model is based on the characteristics of the phase.

Da et al. [49] conducted a heuristic structure based on paraconsistent annotated logic (PAL) capable of simulating the measurement of average values in a dataset and detecting average variations using only PAL concepts. The structure was described as an ANN (PANnetMovAVG) for the retrieval of the moving average. The authors used PANnetMovAVG as an example of its application to assist in the evaluation of an index of the final product value relevant to electrical engineering. Next, testing was carried out using random value information to check PANnetMovAVG's behavior and set the optimum number of methods to create an optimized mathematical structure. The authors then utilized a database with actual electrical voltage measurements produced by an electrical power network from a Brazilian electrical utility grid. In the different tests, PANnetMovAVG correctly observed changes throughout 220-V transmission lines and defined variations in electrical voltage. The results indicate that PANnetMovAVG can be employed to create an efficient architecture for the identification and monitoring of performance scores with applications in different engineering fields, in particular for the identification of the value index in an electricity distribution network.

In Ref. [50], the authors designed and then compared the output of a commonly used ANN with a newly designed ensemble-based algorithm Random Forest (RF). The target application predicted the hourly energy consumption and daylight illuminance values in a classroom in Cardiff, UK. Generally, RF performed much better than ANN in predicting daylight illuminance, with coefficients of determination (R2) of 0.9881 and 0.9799, respectively.

Holkar et al. [51] implemented a neural network that can recognize and thus be trained in finding solutions, recognizing patterns, and classifying information. In this ANN, one can use a control technique that can identify and predict future events. Based on previously obtained samples, the magnitude of control variables is taken, and these measurements are utilized to determine the type of unusual event that may occur on the network. Using the neural network (NN) toolbox in MATLAB, an analysis of the learning and generalization features of elements in the power system is presented.

8.2.5 Construction engineering

Jain and Pathak [52] reviewed a different area of construction engineering where an ANN can be applied as a powerful tool for solving many problems in this field (i.e., construction engineering). ANNs have been utilized for prediction, resources optimization, decision-making, and many more. The researchers also described possible ways for integrating ANNs and expert systems to serve as a basis for developing efficient intelligent systems.

In Ref. [53], the authors used the backpropagation (BP) neural network as a powerful learning ability and adaptability tool in civilian residential quality supervision. An empirical study was carried out based on data from 15 civilian residential projects. Their results showed that the BP neural network model has some limitations, such as the stability of the network being very low. Other methods can be found in Refs. [54–56].

8.2.6 Mechanical engineering

Mechanical engineering is a promising research area in engineering and computer science [54,57–60]. In Ref. [61], finite element software Moldflow, an ANN, and a genetic algorithm were linked to find the optimal gate location within the technical constraints of mold construction. The weld line length was selected as part of the quality index to investigate the effects of gate location on weld line behavior. Moreover, the suggested approach aimed to help engineers in determining other mold design elements such as gate size, gate shape, and coolant systems to compete in terms of quality and cost.

El-hamshary [62] compared the Taguchi method and ANNs, which was carried out to obtain an accurate numerical technique that can predict the surface roughness parameter, Ra, to minimize the cost and time in machining surface operations. The results of this study showed that ANNs are more accurate and precise than the Taguchi method.

Bon and Hui [63] employed ANN to predict the color defects number in furniture manufacturing to improve decision-making. An experiment was conducted by collecting data about color defects from a furniture company, which were analyzed by MATLAB R2015a using a neural network toolbox. The experimental results showed that the NNU helped the company to make the optimal decision based on a series of prediction processes.

In Ref. [64] the author summarized the application of ANNs in mechanical fault diagnosis, mechanical structure analysis, geometry modeling, mechanical design, and also optimization of ANN technology. Different neural network algorithms have been discussed, such as BP neural network, Hopfield neural network, and SOM network. The work

Table 8.1 Summary of the NNA applications.

Field	Method	Description	Problem	References
Chemical		engineering	LM and BFGS	The neural network will help to predict two parameters in safety evaluation
Safety of reactors	Mazrou [34]			
	COD and ANN	Chemical oxygen needs for dairy effluent after electrocoagulation	Wastewater treatment by electrocoagulation	Valente et al. [36]
	HNNRF	Hybrid neural network algorithm with rate function with mechanistic equations	Identify nonlinear behavior	Chang et al. [37]
	NMPC	Continuous reactor for three-level catalytic slurry hydrogenation	Complicated three-level reaction kinetics	Li and Li [38]
Civil		engineering	ANNs	Reviewed the different architectures of ANNs and their learning process

(Continued)

Table 8.1 (Continued)

Field	Method	Description	Problem	References
Learning process	Singh et al. [41]			
	AASM	Apply ANNs to predict the engineering characteristics of concrete	Engineering features	Chen et al. [42]
Computer		engineering	ANN-ML	Study the theoretical and dynamic properties of artificial neural networks
Recognize patterns	Kaur et al. [43]			
	ANN and AI	components	The system uses ANN for the prediction of material removal rate and tool wear	Material removal rate
Katukam [69]				
	ANN-VOA	Developed an intelligent system with the objective of optimizing and improving student's self-learning process and improving student's self-learning during courses	Self-learning	Kose and Arslan [45]
Power and energy	ANN-ESDD	Investigation of the leakage current of an insulator using an ANN	Electrical power transmission	Ananth and Ravindran [70]

	ANNPAL	Simulates the computation of average values of data in a dataset and also detects variations from the average	Electric voltage	da Silva Filho et al. [49]
	ANN-RF	Predicting energy consumption and daylight illuminance values based on a simulation model of a classroom	Energy consumption	Ahmad et al. [50]
	ANN	Fault detection and classification in an electrical power transmission system using an ANN	Electrical power	Holkar and Holkar [71]
Construction		engineering	BPNN	A review paper for adopting neural network algorithms in different construction engineering areas
Decision-making	Jain and Pathak [52]			
	BPNN	Improve methods of quality supervision in civilian residential areas using a BP neural network model	Complex nonlinear relationship	Han and Zhang [53]
Mechanical		engineering	ANN/GA	Find the gate location which leads to a minimum weld line length

(*Continued*)

Table 8.1 (Continued)

Field	Method	Description	Problem	References
Injection mold	Sedighi et al. [61]			
	Taguchi-ANN	Compared the Taguchi method with an ANN in predicting the surface roughness parameter, Ra	Cost and time	El-hamshary [62]
	ANN	Utilizing an ANN to forecast the product defects in furniture	Furniture industry	Bon and Hui [63]
	BPNN	Review paper using an ANN in mechanical engineering	Geometry modeling	Liu [64]
	BPNN	Providing a framework for modeling and predicting a silk wire-rope scaffold in a hierarchical structure using ANN	Wire-rope	Naghashzargar et al. [65]
Geotechnical		engineering	ANN	Providing a review of previous works on ANN applications in geotechnical engineering applications
Classification	Moayedi et al. [66]			
	BPANN	Using BP-ANN observations to forecast future displacements in the region of the tunnel	Guanyin Temple tunnel	Chao et al. [67]
	ANN	Presenting a systematic review for the ANN methodologies and applications in the geotechnical engineering field	Time efficiency	Zhao [68]

highlighted the strengths and features of employing ANNs for solving mechanical engineering problems, as well as providing new opportunities and greater development in future works.

Naghashzargar et al. [65] designed an ANN model using the BP algorithm to predict the mechanical properties of the wire-rope scaffold engineered tendon and ligament. The L32 Taguchi orthogonal matrix has been utilized to design the most effective parameters of the wire-rope scaffold. All results were analyzed based on sensitivity analysis.

8.2.7 Geotechnical engineering

Moayedi et al. [66] provided a review of previous works on ANN applications in geotechnical engineering applications. Different categories of ANN and their applications in the geotechnical engineering field have been explored. The findings of this review study endorsed the use of the ANN approach for classifying oil and rock clusters, as well as forecasting future environmental conditions, including the plastic properties of soil and rock, the stableness of slopes, and subgrade sedimentation.

In Ref. [67], ANN is applied to predict the future situation of the surrounding rock stress change at the Guanyin Temple tunnel. A BP NN was fed with daily continuous observations to forecast the future displacement in the region of the tunnel. The authors concluded that the NN could provide accurate expectations as long as more learning data could be fitted to the network.

Zhao [68] presented a systematic review of the ANN methodologies and applications in the geotechnical engineering field. The reviewed articles were categorized into 17 application areas in the geotechnical engineering field. The author concluded that the NN is characterized by its simplicity in use and time efficiency, which made it a better choice than using traditional mathematical modeling and the finite element method (Table 8.1).

8.3 Conclusion

Recently, ANNs have become the focus of attention for many researchers in various fields, due to the development of data (i.e., big data and cloud computing). The use of ANN aids in improving the traditional methods of finding the optimal solutions, as well as utilization of the BP-ANN to predict efficient solutions with the lowest costs and to avoid potential problems. This work has highlighted using ANN and BP-ANN in different applications in the engineering field. Also, it focused on the benefits of ANNs, which allows it to be used to solve various problems in different fields in the future.

Conflicts of interest

The authors declare that they have no conflicts of interest.

References

[1] R.A. Ibrahim, L. Abualigah, A.A. Ewees, M.A. Al-Qaness, D. Yousri, S. Alshathri, et al., An electric fish-based arithmetic optimization algorithm for feature selection, Entropy 23 (2021) 1189.

[2] L.M.Q. Abualigah, et al., Feature Selection and Enhanced Krill Herd Algorithm for Text Document Clustering, Springer, 2019.

[3] L.M. Abualigah, A.T. Khader, E.S. Hanandeh, A new feature selection method to improve the document clustering using particle swarm optimization algorithm, Journal of Computational Science 25 (2018) 456–466.

[4] M. Shehab, L. Abualigah, M.I. Jarrah, O.A. Alomari, M.S. Daoud, Artificial intelligence in software engineering and inverse, International Journal of Computer Integrated Manufacturing (2020) 1–16.

[5] R. Zheng, H. Jia, L. Abualigah, Q. Liu, S. Wang, An improved arithmetic optimization algorithm with forced switching mechanism for global optimization problems, Mathematical Biosciences and Engineering 19 (2022) 473–512.

[6] L. Abualigah, Group search optimizer: a nature-inspired *meta*-heuristic optimization algorithm with its results, variants, and applications, Neural Computing and Applications (2020) 1–24.

[7] L. Abualigah, Multi-verse optimizer algorithm: a comprehensive survey of its results, variants, and applications, Neural Computing and Applications (2020) 1–21.

[8] M.H. Nadimi-Shahraki, A. Fatahi, H. Zamani, S. Mirjalili, L. Abualigah, An improved moth-flame optimization algorithm with adaptation mechanism to solve numerical and mechanical engineering problems, Entropy 23 (2021) 1637.

[9] M. Shehab, H. Alshawabkah, L. Abualigah, A.-M. Nagham, Enhanced a hybrid moth-flame optimization algorithm using new selection schemes, Engineering with Computers (2020) 1–26.

[10] A. Ziadeh, L. Abualigah, M.A. Elaziz, C.B. Şahin, A.A. Almazroi, M. Omari, Augmented grasshopper optimization algorithm by differential evolution: a power scheduling application in smart homes, Multimedia Tools and Applications (2021) 1–29.

[11] M. Abd Elaziz, L. Abualigah, R.A. Ibrahim, I. Attiya, Iot workflow scheduling using intelligent arithmetic optimization algorithm in fog computing, Computational Intelligence and Neuroscience 2021 (2021).

[12] L. Abualigah, D. Yousri, M. Abd Elaziz, A.A. Ewees, M.A. Al-qaness, A.H. Gandomi, Aquila optimizer: a novel *meta*-heuristic optimization algorithm, Computers & Industrial Engineering 157 (2021) 107250.
[13] L. Abualigah, A. Diabat, S. Mirjalili, M. Abd Elaziz, A.H. Gandomi, The arithmetic optimization algorithm, Computer Methods in Applied Mechanics and Engineering 376 (2021) 113609.
[14] L. Abualigah, M. Abd Elaziz, P. Sumari, Z.W. Geem, A.H. Gandomi, Reptile search algorithm (rsa): A nature-inspired *meta*-heuristic optimizer, Expert Systems with Applications 191 (2022) 116158.
[15] L. Abualigah, A. Diabat, P. Sumari, A.H. Gandomi, Applications, deploy-ments, and integration of internet of drones (iod): a review, IEEE Sensors Journal (2021).
[16] L. Abualigah, A. Diabat, Chaotic binary group search optimizer for feature selection, Expert Systems with Applications (2021) 116368.
[17] M.H. Nadimi-Shahraki, A. Fatahi, H. Zamani, S. Mirjalili, L. Abualigah, M. Abd Elaziz, Migration-based moth-flame optimization algorithm, Processes 9 (2021) 2276.
[18] S.A. Naser, I. Zaqout, M.A. Ghosh, R. Atallah, E. Alajrami, Predicting student performance using artificial neural network: in the faculty of engineering and information technology, International Journal of Hybrid Information Technology 8 (2015) 221–228.
[19] M. Oprea, A. Matei, Applying artificial neural networks in environmental prediction systems, in: Proceedings of the 11th WSEAS International Conference on Automation &# 38; Information, 2010, pp. 110–115.
[20] M. Ushada, T. Okayama, A. Suyantohadi, N. Khuriyati, H. Murase, Daily worker evaluation model for sme-scale food production system using kansei engineering and artificial neural network, Agriculture and Agricultural Science Procedia 3 (2015) 84–88.
[21] M. Ushada, T. Okayama, A. Suyantohadi, N. Khuriyati, H. Murase, Com-bining kansei engineering and artificial neural network to assess worker capacity in small-medium food industry, Proceeding of the Electrical Engineering Computer Science and Informatics 1 (2014) 124–128.
[22] Ö.B. Dinler, C.B. Şahin, L. Abualigah, Comparison of performance of phishing web sites with different deeplearning4j models, Avrupa Bilim ve Teknoloji Dergisi (2021) 425–431.
[23] R.A. Zitar, L. Abualigah, N.A. Al-Dmour, Review and analysis for the red deer algorithm, Journal of Ambient Intelligence and Humanized Computing (2021) 1–11.
[24] R.A. Zitar, L. Abualigah, Application of red deer algorithm in optimizing complex functions, 2021 14th International Congress on Image and Signal Processing, BioMedical Engineering and Informatics (CISP-BMEI), IEEE, 2021, pp. 1–6.
[25] E.H. Houssein, M. Dirar, L. Abualigah, W.M. Mohamed, An efficient equilibrium optimizer with support vector regression for stock market prediction, Neural Computing and Applications (2021) 1–36.
[26] S. Baluja, D. Pomerleau, Non-intrusive gaze tracking using artificial neural networks, Advances in Neural Information Processing Systems, The MIT Press, 1994, pp. 753–760.
[27] J. Sietsma, R.J. Dow, Creating artificial neural networks that generalize, Neural Networks 4 (1991) 67–79.
[28] G. Bolt, Investigating fault tolerance in artificial neural networks (1991).
[29] N.S. El Jerjawi, S.S. Abu-Naser, Diabetes prediction using artificial neural network (2018).

[30] L. Abualigah, A. Diabat, A comprehensive survey of the grasshopper optimization algorithm: results, variants, and applications, Neural Computing and Applications (2020) 1–24.
[31] S. Wang, H. Jia, L. Abualigah, Q. Liu, R. Zheng, An improved hybrid aquila optimizer and harris hawks algorithm for solving industrial engineering optimization problems, Processes 9 (2021) 1551.
[32] R. Zheng, H. Jia, L. Abualigah, Q. Liu, S. Wang, Deep ensemble of slime mold algorithm and arithmetic optimization algorithm for global optimiza-tion, Processes 9 (2021) 1774.
[33] S. Wang, Q. Liu, Y. Liu, H. Jia, L. Abualigah, R. Zheng, et al., A hybrid ssa and sma with mutation opposition-based learning for constrained engineering problems, Computational Intelligence and Neuroscience 2021 (2021).
[34] H. Mazrou, Performance improvement of artificial neural networks designed for safety key parameters prediction in nuclear research reactors, Nuclear Engineering and Design 239 (2009) 1901–1910.
[35] A.M. Altabeeb, A.M. Mohsen, L. Abualigah, A. Ghallab, Solving capacitated vehicle routing problem using cooperative firefly algorithm, Applied Soft Computing 108 (2021) 107403.
[36] G. Valente, R. Mendoụca, J. Pereira, L. Felix, Artificial neural network prediction of chemical oxygen demand in dairy industry effluent treated by electrocoagulation, Separation and Purification Technology 132 (2014) 627–633.
[37] J.-S. Chang, S.-C. Lu, Y.-L. Chiu, Dynamic modeling of batch polymerization reactors via the hybrid neural-network rate-function approach, Chemical Engineering Journal 130 (2007) 19–28.
[38] S. Li, Y.-Y. Li, Neural network based nonlinear model predictive control for an intensified continuous reactor, Chemical Engineering and Processing: Process Intensification 96 (2015) 14–27.
[39] M.H. Hassan, S. Kamel, L. Abualigah, A. Eid, Development and application of slime mould algorithm for optimal economic emission dispatch, Expert Systems with Applications 182 (2021) 115205.
[40] A. Eid, S. Kamel, L. Abualigah, Marine predators algorithm for optimal allocation of active and reactive power resources in distribution networks, Neural Computing and Applications (2021) 1–29.
[41] V. Singh, S. Bano, A.K.Y. Sabih Ahmad Dr., Feasibility of artificial neural network in civil engineering, International Journal of Trend in Scientific Research and Development 3 (2019) 724–728.
[42] B.-T. Chen, H.-Y. Wang, C.-C. Wang, J.-N. Chang, Prediction of engineering properties of alkali-activated slag mortar (aasm) using artificial neural network (ann), in: The 2017 World Congress on Advances in Structural Engineering and Mechanics, 2017.
[43] R. Kaur, M. Thapa, B. Sharma, Study of theoratical properties of artificial neural network, a kind of application used in computer science and electronics engineering, International Journal of Science, Engineering and Computer Technology 4 (2014) 82–84.
[44] R. Katukam, P. Behera, Engineering optimization using artificial neural network, International Journal of Innovations in Engineering and Technology (IJIET) 4 (2014) 50–63.
[45] U. Kose, A. Arslan, Optimization of self-learning in computer engineering courses: an intelligent software system supported by artificial neural network and vortex optimization algorithm, Computer Applications in Engineering Education 25 (2017) 142–156.

[46] D. Yousri, M. Abd Elaziz, D. Oliva, L. Abualigah, M.A. Al-qaness, A.A. Ewees, Reliable applied objective for identifying simple and detailed pho-tovoltaic models using modern metaheuristics: comparative study, Energy Conversion and Management 223 (2020) 113279.

[47] L. Abualigah, M. Shehab, M. Alshinwan, S. Mirjalili, M. Abd Elaziz, Ant lion optimizer: a comprehensive survey of its variants and applications, Archives of Computational Methods in Engineering (2020).

[48] A. Ananth, M. Ravindran, Investigation of leakage current of insulator using artificial neural network, International Journal of Engineering Sciences and Research Technology Investigation of Leakage Current of Insulator Using Artificial Neural Network 5 (2016) 667−672.

[49] J.I. da Silva Filho, C.M. Da Cruz, A. Rocco, D.V. Garcia, L.F.P. Ferrara, A.S. Onuki, et al., Paraconsistent artificial neural network for structuring statistical process control in electrical engineering, Towards Paraconsistent Engineering, Springer, 2016, pp. 77−102.

[50] M. Ahmad, J.-L. Hippolyte, M. Mourshed, Y. Rezgui, Random forests and artificial neural network for predicting daylight illuminance and energy consumption (2017).

[51] J. Holkar, V. Fulmali, V. Holkar, Fault analysis on transmission lines using artificial neural network, International Journal of Engineering Sciences and Research Technology 5 (2016) 51−59.

[52] M. Jain, K. Pathak, Applications of artificial neural network in construction engineering and management—a review, International Journal of Engineering Technology, Management and Applied Sciences 2 (2014) 134−142.

[53] G.B. Han, L.N. Zhang, Research on construction engineering quality supervision based on artificial neural network, Advanced Materials Research, Vol. 919, Trans Tech Publ, 2014, pp. 1471−1476.

[54] M. Abd Elaziz, A. Dahou, L. Abualigah, L. Yu, M. Alshinwan, A.M. Khasawneh, et al., Advanced metaheuristic optimization techniques in applications of deep neural networks: a review, Neural Computing and Applications (2021) 1−21.

[55] P. Sumari, S.J. Syed, L. Abualigah, A novel deep learning pipeline architecture based on cnn to detect covid-19 in chest x-ray images, Turkish Journal of Computer and Mathematics Education (Turcomat) 12 (2021) 2001−2011.

[56] M. Safaldin, M. Otair, L. Abualigah, Improved binary gray wolf optimizer and svm for intrusion detection system in wireless sensor networks, Journal of Ambient Intelligence and Humanized Computing 12 (2021) 1559−1576.

[57] M. Shehab, L. Abualigah, H. Al Hamad, H. Alabool, M. Alshinwan, A.M. Khasawneh, Moth−flame optimization algorithm: variants and applications, Neural Computing and Applications 32 (2020) 9859−9884.

[58] L. Abualigah, M. Shehab, A. Diabat, A. Abraham, Selection scheme sensitivity for a hybrid salp swarm algorithm: analysis and applications, Engineering with Computers (2020) 1−27.

[59] F. Gul, I. Mir, L. Abualigah, P. Sumari, Multi-robot space exploration: an augmented arithmetic approach, IEEE Access 9 (2021) 107738−107750.

[60] C.B. Şahin, L. Abualigah, A novel deep learning-based feature selection model for improving the static analysis of vulnerability detection, Neural Computing and Applications (2021) 1−19.

[61] R. Sedighi, M.S. Meiabadi, M. Sedighi, Optimisation of gate location based on weld line in plastic injection moulding using computer-aided engineering, artificial neural network, and genetic algorithm, International Journal of Automotive & Mechanical Engineering 14 (2017).

[62] N.A. El-hamshary, International journal of engineering sciences & research technology numerical prediction of surface roughness in turning operation using taguchi method and artificial neural network, 4 (2015).
[63] A.T. Bon, H.S. Hui, Industrial engineering solution in the industry: artificial neural network forecasting approach, International Conference on Industrial Engineering and Operations Management Rabat (2017).
[64] Y. Liu, The review of intelligent mechanical engineering based on artificial neural network, in: 2015 International Conference on Intelligent Systems Research and Mechatronics Engineering, Atlantis Press, 2015.
[65] E. Naghashzargar, D. Semnani, S. Karbasi, Improving the mechanical properties of wire-rope silk scaffold by artificial neural network in tendon and ligament tissue engineering, Journal of Engineered Fibers and Fabrics 10 (2015) 94–102.
[66] H. Moayedi, M. Mosallanezhad, A.S.A. Rashid, W.A.W. Jusoh, M.A. Muazu, A systematic review and *meta*-analysis of artificial neural network application in geotechnical engineering: theory and applications, Neural Computing and Applications (2018) 1–24.
[67] Z. Chao, G. Ma, Y. Zhang, Y. Zhu, H. Hu, The application of artificial neural network in geotechnical engineering, IOP Conference Series: Earth and Environmental Science, Vol. 189, IOP Publishing, 2018, p. 022054.
[68] N. Zhao, The application of bp artificial neural network in geotechnical engineering, Applied Mechanics and Materials, Vol. 580, Trans Tech Publ, 2014, pp. 823–826.
[69] R. Katukam, Engineering optimization using artificial neural network, International Journal of Innovations in Engineering and Technology (IJIET), 4, 2014, pp. 50–63.
[70] A. Ananth, M. Ravindran, Investigation of leakage current of insulator using artificial neural network, International Journal of Engineering Sciences and Research Technology 5 (2016) 667–672.
[71] F.V. Holkar, Jayati, V. Holkar, Fault analysis on transmission lines using artificial neural network, International Journal of EngineerIng Sciences and Research Technology 5 (2016) 51–59.

CHAPTER NINE

Incremental deep learning model for plant leaf diseases detection

Salima Ouadfel[1], Wafa Mousser[2], Ismail Ghoul[3] and Abdelmalik Taleb-Ahmed[4]

[1]Department of Computer Sciences and Applications, New Information and Communication Technologies Faculty, Abdelhamid Mehri Constantine 2 University, Constantine, Algeria
[2]Department of Computer Sciences and Applications, New Information and Communication Technologies Faculty, Laboratory of Complex Systems' Modeling and Implementation, Abdelhamid Mehri Constantine 2 University, National Biotechnology Research Center, Constantine, Algeria
[3]Department of Computer Sciences and Applications, New Information and Communication Technologies Faculty, Laboratory of Complex Systems' Modeling and Implementation, Abdelhamid Mehri Constantine 2 University, Constantine, Algeria
[4]Institut d'Electronique de Microélectronique et de Nanotechnologie (IEMN), UMR 8520, Université Polytechnique Hauts de France, Université de Lille, CNRS, Valenciennes, France

Contents

9.1 Introduction

Good agricultural management aims to improve and increase agricultural productivity without altering natural resources and ensuring a good quality of life for the farmer. However, this goal is threatened by certain diseases that can affect crop productivity. This threat is of constant concern to

Artificial Neural Networks for Renewable Energy Systems and Real-World Applications.
DOI: https://doi.org/10.1016/B978-0-12-820793-2.00011-2

farmers. More specifically, they are looking to find a way that improves crop management and reduces the negative consequences of fertilizers and pesticides [1]. Being able to detect plant diseases early and effectively has become a major challenge in recent years. In order to be able to make rapid and efficient diagnoses of plant diseases, the phytopathologist must be able to detect the various characteristic symptoms [2]. With recent advances in computer vision, plant protection has become more precise and has broadened the field of application of computer vision in agriculture. Several approaches have been proposed in the literature to detect plant diseases, and among these are deep neural networks [3]. Neural networks are inspired from biological nervous systems, and have been used with great success in engineering applications, including many computer vision applications. The popularity of deep learning is due to its ability to learn and represent data at different scales or levels of abstraction using multiple layers of neurons. This growing interest is mainly due to the abundance of complex data coming from different sources.

9.2 Related works

In recent years, deep learning has been used extensively in various fields of application. In particular, its great success in the recognition and classification of images has motivated the authors to use it in the identification and classification of plant diseases in the field of agriculture [4]. In Ref. [5] the authors combine a modified version of convolutional neural network (CNN) with the Random Forest (RF) classifier for the classification of leaves. They evaluated their model using the classification accuracy, which reached 97.3%. In Ref. [6], another work was proposed based on the AlexNet model. The authors in Ref. [7] proposed the use of deep learning models for the identification of plants and achieved a success rate of 91.78%. In Ref. [8], the authors introduce a CNN model for the classification of diseases in maize plants. Three CNN models, namely AlexNet, GoogLeNet, and ResNet, were proposed in Ref. [9] to identify tomato leaf diseases. Experimental results revealed the superiority of ResNet over the two other models. LeNet was applied in Ref. [10] to recognize diseases in banana leaf on both a color and gray scale. In Ref. [11], popular deep learning models like GoogLeNet, ResNet-50, ResNet-101, and Inception-v3 were combined with support vector machine (SVM), K-nearest neighbors, and extreme learning machine for the classification of eight different plant diseases. The combination of ResNet-50 with the SVM classifier gave better performance. In Ref. [12] the authors used AlexNet and GoogLeNet CNN models for classification of plant diseases using the PlantVillage dataset. From the experimental results, they observed the superiority of GoogLeNet. In another work in Ref. [13], the authors compared AlexNet and GoogLeNet models for tomato plant diseases, and again GoogLeNet outperformed AlexNet. In Ref. [14] the authors proposed a deep learning model to identify rice plant diseases.

The authors in Ref. [15] proposed the LwF: learning without forgetting algorithm which is able to update its output by learning new classes without access to the old datasets. It performs well for both new and old tasks. More recently, the authors of Ref. [16] proposed iCaRL: an incremental classifier that relies on a set of exemplar samples from previously learnt classes while learning new ones. The classification is ensured by a *nearest-mean-of-exemplars* which computes the feature-vector for a given input and returns the label of the most similar average-feature vector of exemplar sets.

In this work, deep learning and incremental classification technics are combined to propose an incremental deep learning framework for plant leaf diseases detection.

9.3 Proposed approach

The core idea of our methods is to incrementally train a deep learning model using new classes of data combined with selected classes from old data. These selected data allow knowledge of previous tasks to be protected while new tasks are learnt. As shown in Fig. 9.1, for the first task t_1 we train and test the network for D_1 ($Dataset_1$), followed by t_2 when D_2 is available. The network is retrained on D_2 with samples from D_1, where D_1 (t_2) is the selected samples from D_1 (t_2) as a review data for t_2. However, the old test data remain constant on all tasks and are not reduced. Note that the new task should contain one or more classes. The equations below explain how the general methods based on review data work:

- $training(t_3) = train[D_1(t_3)] \cup train[D_2(t_3)] \cup train[D_3]$
- $testing(t_3) = test[D_1] \cup test[D_2] \cup test[D_3]$

Fig. 9.2 shows the two main components of our approach. The first one is the *deep learning model* and the second is the *DataSelector* and *Memory*. The former contains a ConvNets as a feature extractor and an

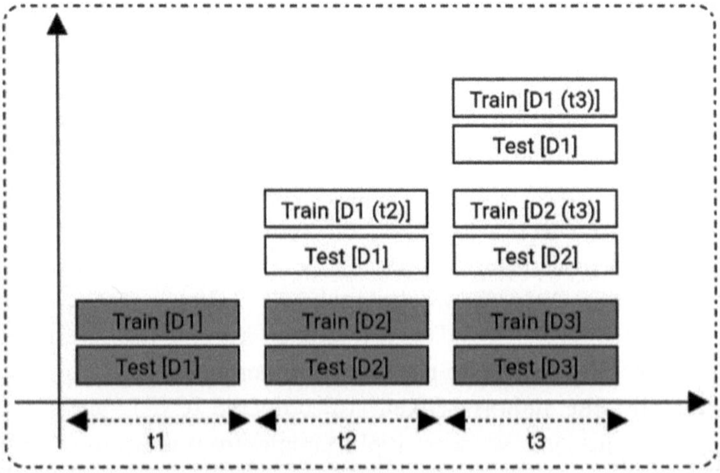

Figure 9.1 Incremental training for three tasks.

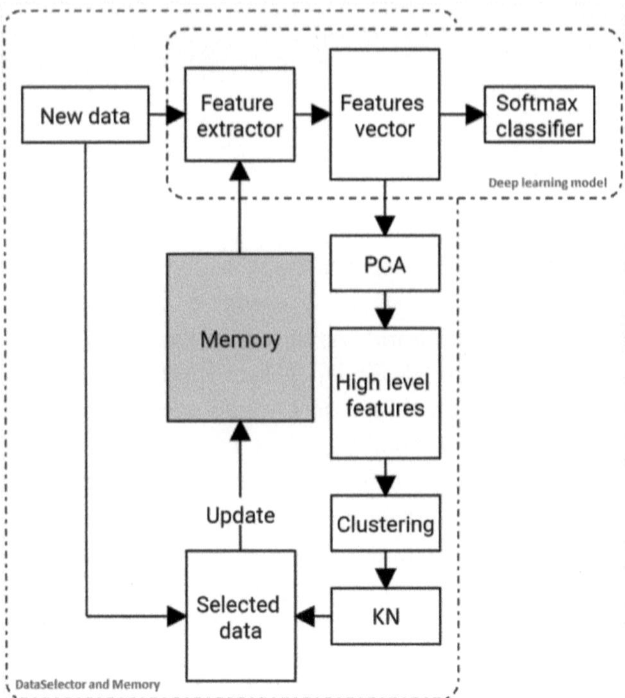

Figure 9.2 Our approach architecture.

fully connected for the classification task. The latter is used to select data from each new class and support old tasks when the model is trained on the new ones. Each component of the proposed approach is detailed in the following sections.

9.3.1 The deep learning model

As mentioned before, the *deep learning model* of the proposed framework is divided into two parts: the *feature-extractor* that maps an image into a *d*-dimensional feature vector and the *classifier*.

In order to learn incrementally, the number of classes the model outputs increases at each task. The goal is to allow continual learning of new data starting from the old trained model, commonly known as *fine-tuning* [1,2]. Fig. 9.3 shows how the model is built and prepared for a new task.

In principle, we use transfer learning techniques but without freezing the weights of the base network. As can be seen in Fig. 9.3, the base network's (trained model on old tasks) parameters are untouched, and the newly added n nodes are fully connected to the layer before, where n is the number of new classes. Weights corresponding to these n added nodes are then initialized by θ. This is done because we assume that we start from a trained network on old tasks to preserve the same performance before training on a new task.

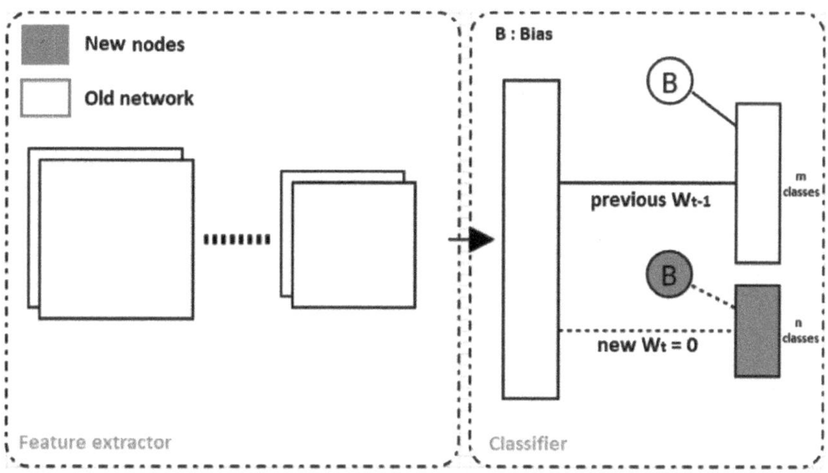

Figure 9.3 The deep learning model.

9.3.2 DataSelector and Memory

As mentioned before, the *DataSelector* is used to select data samples from each new class as a review data. The selected data are then maintained in the *Memory* component with the old selected one. The main question is how these data could be selected?

Our approach is based on data diversity, that is similar data are grouped into the same group, as shown in Fig. 9.4A. Note that similar data have close feature vectors. After that, the data size is reduced by choosing fewer samples from each group. As shown in Fig. 9.4B, samples that have not been chosen are close to those that have been chosen. This means that the features' diversity is preserved. This *DataSelector* works on each new class separately.

9.3.2.1 Feature-extractor

The convolutional layer stack of the previously trained deep model represents the *feature-extractor* module of our approach. It is used to compute the features for new task images. The size of the obtained features is equal to VxS, where V is the number of entered images, and S is the size of one image feature's vector. Note that any other efficient method such as a pretrained CNN can be used as a *feature-extractor* module.

9.3.2.2 Principal component analysis

Principal component analysis (PCA) is the most popular technology for reducing dimensions or datasets while maintaining their diversity. The projection matrix resulting from PCA can be seen as a change of coordinates to a coordinate system where the coordinates are in descending order of importance [2]. Our method has two scenarios:

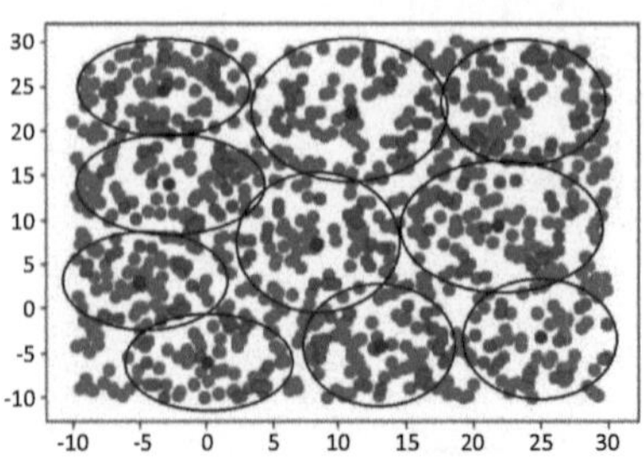

(A) KN for 10 clusters

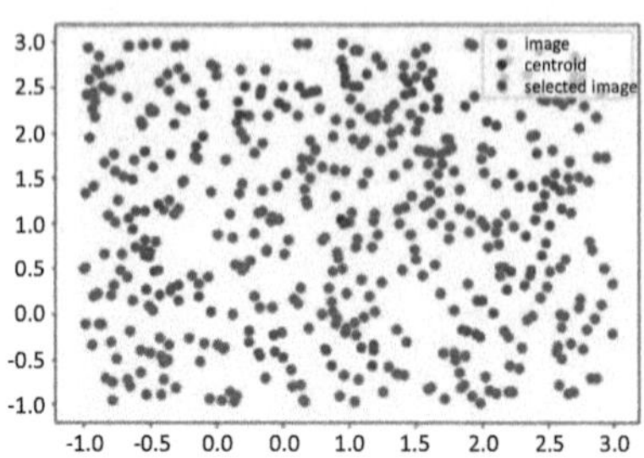

(B) KN process result for one cluster

Figure 9.4 K-neighbors process results (in two dimensions). (A) KN results for 10 clusters, (B) KN results for one cluster.

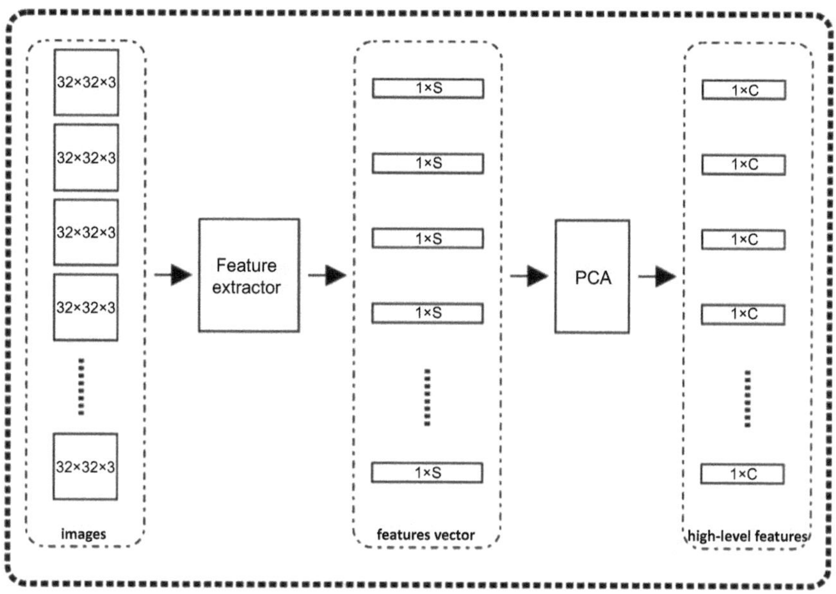

Figure 9.5 Extract high-level features from images by FE and PCA.

- First, PCA is applied to the *feature-extractor* outputs (feature vectors) to reduce the dimensions of each feature vector from S to C depending on the variance ratio value chosen (see Fig. 9.5).

  The PCA takes as input a VxS matrix, where V is the number of samples (rows) and S is the number of features for each sample (columns). The PCA returns then a VxC matrix, where each column is orthogonal to the other columns. These columns identify the axes that account for the largest amount of variance in the features called *principal components.* The PCA helps to select variables according to the magnitude (from largest to smallest in absolute values) of their coefficients. In order to preserve a large part of the features variance, the minimum number C of dimensions is computed, which is required to preserve Z% of features variance defined in ratio variance matrix (i.e., $Z=99$).
- Second, PCA is used to determine the number of clusters. Eq. (9.1) shows how to compute it. Note that if $N=CxS$ is close to 1 (i.e., there is much variation in data), then more clusters will be present.

$$N_c = \frac{C * NC_{max}}{S} \tag{9.1}$$

where NC_{max} is the maximum number of clusters, which is set equal to 100 by default and may be changed.

9.3.2.3 Clustering

The main idea of the clustering algorithm is to use the high-level features, extracted from the input images, as training samples for the *K*-mean algorithm and obtain images grouped by their similarity (see Algorithm 9.1).

Algorithm 9.1: Clustering algorithm

Data: $\mathbf{F} = \{f_1, \ldots, f_v\}$ high-level features of all images
require n number of clusters;
Result: $X = X_1, \ldots, X_n$ classified samples by number of clusters
Begin
$X = K\text{-means}(n, F)$ classified samples by n clusters
End

9.3.2.4 K-neighbors

In order to allow our framework to choose *K* samples from a class efficiently, where *K* is the target samples per class, Algorithm 9.2 is proposed. This is based on unsupervised nearest neighbors (from the Sklearn library), and uses *Euclidean distance* to find the neighbors between two datasets. This algorithm is used to find the distances between a cluster centroid and its items.

Algorithm 9.2: K-neighbors

Data: $\boldsymbol{C} = \{\boldsymbol{C_1}, \ldots, \boldsymbol{C_r}\}$ Current centroid sets
require $\boldsymbol{X} = \{\boldsymbol{X_1}, \ldots, \boldsymbol{X_n}\}$ High-level features vector classified by clustering
require $\boldsymbol{K} = \{\boldsymbol{K_1}, \ldots, \boldsymbol{K_n}\}$ Number of target samples per cluster
Result: neighbors_indices Indices of target samples
Begin
For C_i in C **Do**
D, indices = NearestNeighbors(C, X_i, indices)
$Avg = sumDq$
neighbors_indices + = close_to_Avg(D,K_i,indices)
End For
End

After that, from each $cluster_i$, the green K_i samples are chosen by selecting the closest neighbors by a threshold. The last is the average of all the distances between a centroid and its item's *Avg* as shown in Fig. 9.4A and B.

Algorithm 9.2 shows how to detect these efficient samples, and the obtained 2D results are shown in Fig. 9.4.

$$P = \frac{K}{\text{number of all training classes images}} \tag{9.2}$$

$$K_i = (\text{number of \_items} \in \text{cluster}_i) * P \tag{9.3}$$

where P is the percentage of selected data, and $K \geq \sum_{i=1}^{n} K_i$.

9.3.2.5 Selected data

Once the neighbor's vector indices from K-neighbors (*KN*) process are detected, its corresponding original data can be found, and is used to construct and update the *Memory*.

Also, according to the clusters' output of each class, a record of the selected data is defined. For example, if a record $r = [W, V]$, this means that W is the size of the selected data from the first cluster, while V is the size of the selected data from the second one.

9.3.2.6 Memory

Whenever our framework encounters new classes, the *Memory* component is updated (Fig. 9.6). To do this, all classes are treated equally. Inspired by the iCaRL [16] algorithm, if T is the total number of current classes and M is the memory size available for our framework, then, for each class, the number of samples is obtained by $K = \frac{M}{T}$.

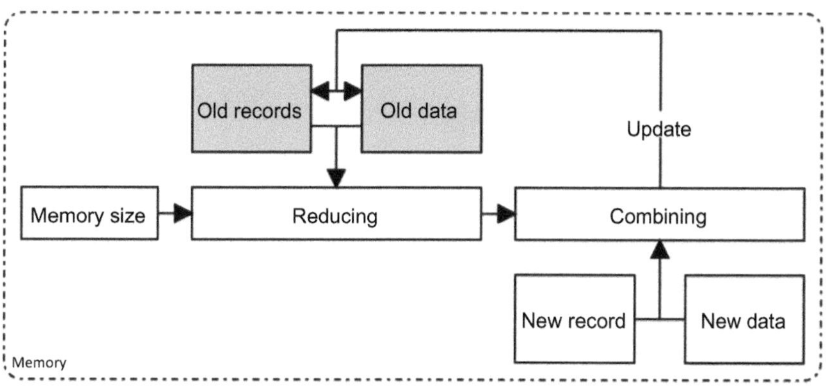

Figure 9.6 Memory components.

When the framework attempts to update the memory by reducing the selected data for each old class, samples are removed from the memory by maintaining the data diversity. This is achieved by a proposed record, which contains a series of selected data indices sorted according to their classification by clustering, where each series represents a group of similar images, that is, if P% of selected data from old class *C* is to be removed, then P% is removed from each series.

The proposed approach is a machine learning hybridization of supervised and unsupervised algorithms allowing learning new tasks while retaining knowledge of previous ones, and thus avoiding catastrophic forgetting.

9.4 Experimental results

In this section, the proposed approach and previously presented data are experimented on for plant diseases classification. Different experiments and comparisons are carried out as detailed in the following.

9.4.1 Plant diseases dataset

The PlantVillage Dataset [17] was used to test our framework for the plant diseases detection classification. The cited dataset contains 326.38 MB of 15 disease classes. From this 20% are used for testing, while the remaining 80% represent the training data. The dataset is organized into two folders, *train* and *test*, each of which contains 15 subfolders for different classes. In order to ensure fair comparisons, this configuration is kept for all the experiments. Fig. 9.7 shows the input distribution for our experiments.

9.4.2 The model's architecture

To achieve our experiments, a customized deep neural network model is used. On one hand, the *feature-extractor* module is composed of three *ConvNets* blocks followed by a *Dropout* layer, as shown in Fig. 9.8. Each of the three blocks contains two *Conv2D* layers and a *Max-pooling* layer. On the other hand, regarding the classification part, the feature vector is flatten and fed into a neural network consisting of 512 neurons with *ReLUs* and *dropout* regularizer and followed by a *Softmax* output layer of *C* units.

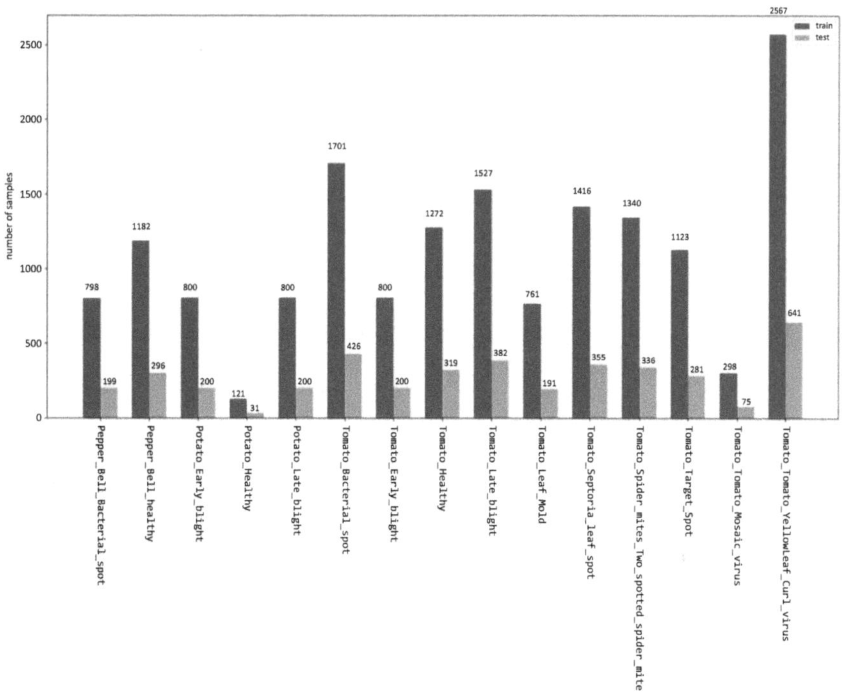

Figure 9.7 Experimental statistical details of the PlantVillage Dataset.

Layers	Parameters
Data (input)	Size: 32×32×3
Conv1/ 3×3 (convolutional)	Pad: 1, stride: 1, filters : 32
Conv2/ 3×3 (convolutional)	Pad: 1, stride: 1, filters : 32
Max_pooling1 (pooling)	Kernel: 2
Dropout	0.2
Conv3/ 3×3 (convolutional)	Pad: 1, stride: 1, filters : 64
Conv4/ 3×3 (convolutional)	Pad: 1, stride: 1, filters : 64
Max_pooling2 (pooling)	Kernel: 2
Dropout	0.3
Conv5/ 3×3 (convolutional)	Pad: 1, strid: 1, filters : 128
Conv6/ 3×3 (convolutional)	Pad: 1, stride: 1, filters : 128
Max_pooling3 (pooling)	Kernel: 2
Dropout	0.4
Flatten	
FC1 (fully connected)	Neurons: 512
Dropout	0.5
FC2 (output)	Neurons: (C classes)

Figure 9.8 The deep model summary.

9.4.3 Hyperparameters and evaluation

Once our deep model was built, the *Adam*'s optimizer was used in combination with *categorical cross-entropy* loss function for compilation. The *accuracy*

Parameters	Values	Description
Initial-lr	0.001	Learning rate used when the network is training on only the first task's dataset.
Later-lr	0.0001	Learning rate used when the network is being trained on any later task.
minibatch-size	32	The number of items trained from during each mini-batch.
Patience	8	Training is stopped when the network has not improved in its validation accuracy for this number of epochs.
Epochs	50	The number of epochs to train the model. An epoch is an iteration over the set of X and Y.

Figure 9.9 The deep model's hyperparameters.

measure was used to evaluate our model. Fig. 9.9 summarizes the hyperparameter values used to conduct our experiments. The test partition for the new and old tasks was used to perform the final evaluation of our approach.

9.4.4 Influence of memory size and batches

As mentioned before, our method uses a limited size memory. The proposed approach is tested using different values for batches and for the memory size. Obtained results are presented in this section. Fig. 9.10 shows the final validation accuracies obtained for each experiment. Where, in this series of experiments, three memories of size 500, 1000, and 2000 are tested. Each of these memories represents the number of examples of old data preserved in the review process. However, for the task batches, three different sizes of 1, 3, and 5 are used to indicate the number of new classes present per batch.

In order to highlight our incremental approach, our deep learning network is trained from scratch for multiclass classification, using all available data (new and old) as an *upper-bound*. The obtained results are presented in the *all_data* line of Fig. 9.10. The challenge is then to achieve close results to those previously obtained. The graphs in Fig. 9.11 show the accuracy variation while learning different tasks.

On one hand, from Fig. 9.11, we can see that the *2000 examples* in memory version perform the best compared to the *1000 and 500 examples* version for the final evaluation, which is explained by the fact of preserving a lot of examples in memory.

On the other hand, the increase of the memory leads to an increase in the training time, giving better results (Fig. 9.12). However, when a small memory value is used, the learning time is reduced and the results are also reduced. Fig. 9.13 shows the obtained results using an Intel i5-6200 CPU with 8 GB of RAM. Apparently, the precision increases with the memory increasing. We observe also that, when the number of classes increases per batch, the performance increases faster than the batch which contains less

Classes per batch	Memory size	Average of 5 accuracies	Best accuracy
1	2000	0.878	0.882
3	2000	0.900	0.916
5	2000	0.917	0.926
1	1000	0.834	0.843
3	1000	0.846	0.868
5	1000	0.882	0.889
1	500	0.759	0.768
3	500	0.790	0.803
5	500	0.846	0.860
All data	/	/	**0.94**

Figure 9.10 Five accuracies average results of different sizes for memory and batch.

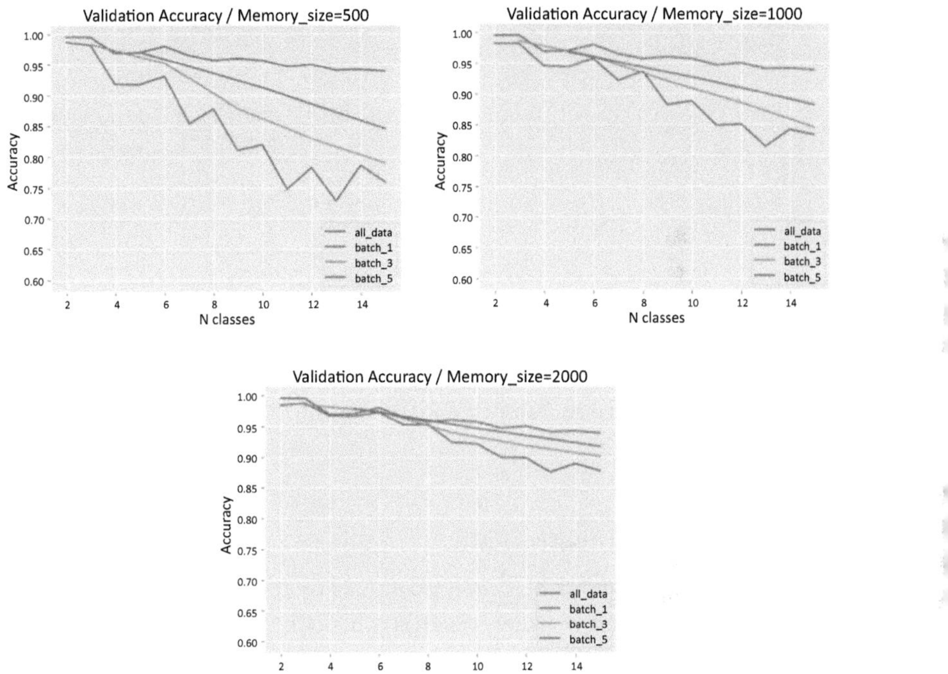

Figure 9.11 Average of training accuracies for different batches and different memory sizes, and *all_data*.

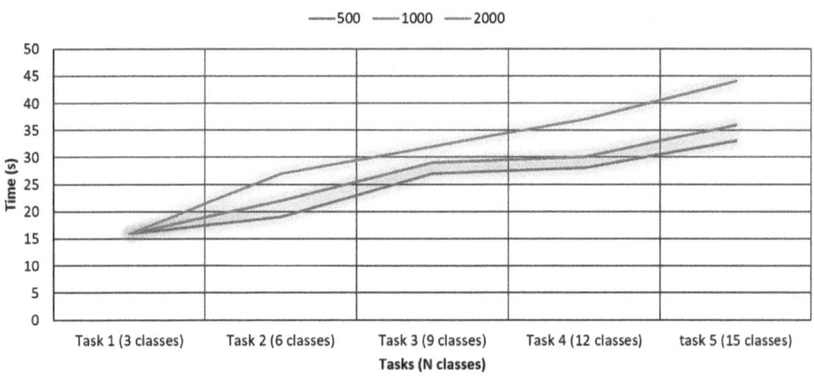

Figure 9.12 Epoch time required for various memory sizes and *batch_size* = 3.

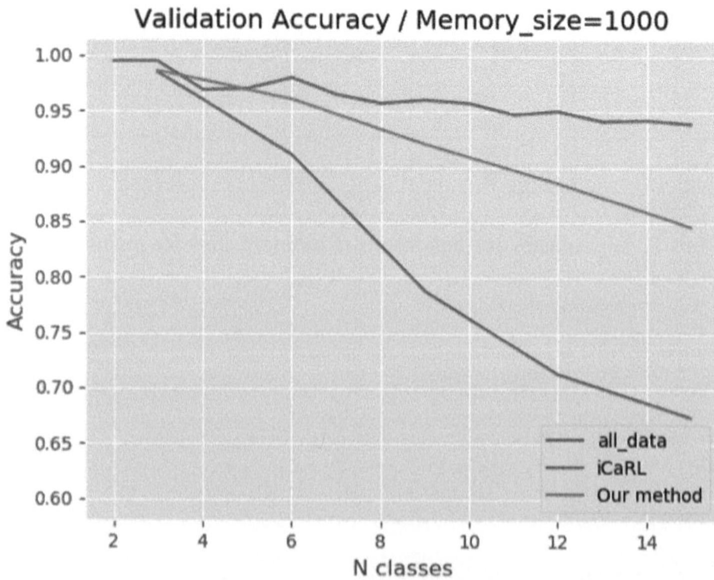

Figure 9.13 Comparing five average validation accuracies on PlantVillage with three classes per batch.

classes, because the large batches contain the most training data that solve the overfitting problem.

9.4.5 Comparison with iCaRL

In this subsection, we make a comparison between our approach and the iCaRL method which is widely used as a benchmark for incremental deep learning. For this, we modify some parameters in our algorithms as follows:

- For a fair comparison with iCaRL and our method, we use the same networks and follow the same class division protocols.
- For each new observed class, iCaRL uses a nearest-mean-of-exemplars which calculates the mean features vector (MFV) of all the training examples for a class, then it chooses the examples close to this MFV [16]. In order to make our method similar to iCaRL, we removed the PCA process and changed some parameters in the clustering algorithm and KN process. First, with a fixed number of clusters, which is at all times equal to 1, we consider its centroid as an MFV. Second, we changed the KN process to detect the examples closest to the centroid. Here, the clustering algorithm and the new KN process do the same job as the nearest-mean-of-exemplars process.

As shown in Fig. 9.13, our proposed framework performs better than iCaRL by 20% when the 15 classes are incrementally learnt. The difference is less important before the learning of the first six classes. This is due to the fact that iCaRL does not support data diversity and is overfitted on the selected exemplar-set with an increase in the number of classes.

9.5 Conclusion

In this work, we have proposed an incremental learning method based on deep neural networks for the classification of plant diseases. The goal of our approach is to treat the catastrophic forgetting of CNNs by adapting the model to the arrival of new data without forgetting its previously existing knowledge. Different experiments and comparisons are carried on PlantVillage dataset. The experimental results have shown that our method is efficient and provides better results than the incremental model iCaRL.

References

[1] J.A. Lucas, Advances in plant disease and pest management, The Journal of Agricultural Science 149 (S1) (2011) 91.
[2] M.B. Riley, M.R. Williamson, O. Maloy, Plant disease diagnosis, The Plant Health Instructor 10 (2002).
[3] P.M. Buscema, G. Massini, M. Breda, W.A. Lodwick, F. Newman, M. Asadi-Zeydabadi, Artificial neural networks, Artificial Adaptive Systems Using Auto Contractive Maps, Springer, 2018, pp. 11–35.
[4] A. Kamilaris, F.X. Prenafeta-Boldú, Deep learning in agriculture: a survey, Computers and Electronics in Agriculture 147 (2018) 70–90.
[5] D. Hall, C. McCool, F. Dayoub, N. Sunderhauf, B. Upcroft, Evaluation of features for leaf classification in challenging con- ditions, 2015 IEEE Winter Conference on Applications of Computer Vision, IEEE, 2015, pp. 797–804.
[6] S.H. Lee, C.S. Chan, P. Wilkin, P. Remagnino, Deep-plant: plant identification with convolutional neural networks, 2015 IEEE International Conference on Image Processing (ICIP), IEEE, 2015, pp. 452–456.
[7] Y. Sun, Y. Liu, G. Wang, H. Zhang, Deep learning for plant identification in natural environment, Computational Intelligence and Neuroscience (2017) 2017.
[8] M. Sibiya, M. Sumbwanyambe, A computational procedure for the recognition and classification of maize leaf diseases out of healthy leaves using convolutional neural networks, AgriEngineering 1 (1) (2019) 119–131.
[9] K. Zhang, Q. Wu, A. Liu, X. Meng, Can deep learning identify tomato leaf disease? Advances in Multimedia (2018) 2018.
[10] J. Amara, B. Bouaziz, A. Algergawy, A deep learning- based approach for banana leaf diseases classification, Datenbanksysteme für Business, Technologie und Web (BTW 2017)-Workshopband (2017).

[11] M. Türkoğlu, D. Hanbay, Plant disease and pest detection using deep learning-based features, Turkish Journal of Electrical Engineering & Computer Sciences 27 (3) (2019) 1636–1651.
[12] S.P. Mohanty, D.P. Hughes, M. Salathé, Using deep learning for image-based plant disease detection, Frontiers in Plant Science 7 (2016) 1419.
[13] S. Wallelign, M. Polceanu, C. Buche, Soybean plant disease identification using convolutional neural network, in: The Thirty-First International Flairs Conference, 2018.
[14] Y. Lu, S. Yi, N. Zeng, Y. Liu, Y. Zhang, Identification of rice diseases using deep convolutional neural networks, Neurocomputing 267 (2017) 378–384.
[15] Z. Li, D. Hoiem, Learning without forgetting, IEEE Transactions on Pattern Analysis and Machine Intelligence 40 (12) (2017) 2935–2947.
[16] S.-A. Rebuffi, A. Kolesnikov, G. Sperl, C.H. Lampert, iCaRL: incremental classifier and representation learning, Proceedings of the IEEE Conference on Computer Vision and Pattern Recognition, IEEE, 2017, pp. 2001–2010.
[17] Kaggle, Plantvillage dataset, 2018.

CHAPTER TEN

Incremental learning of convolutional neural networks in bioinformatics

Wafa Mousser[1], Salima Ouadfel[2] and Abdelmalik Taleb-Ahmed[3]

[1]Department of Computer Sciences and Applications, New Information and Communication Technologies Faculty, Laboratory of Complex Systems' Modeling and Implementation, Abdelhamid Mehri Constantine 2 University, National Biotechnology Research Center, Constantine, Algeria

[2]Department of Computer Sciences and Applications, New Information and Communication Technologies Faculty, Abdelhamid Mehri Constantine 2 University, Constantine, Algeria

[3]Institut d'Electronique de Microélectronique et de Nanotechnologie (IEMN), UMR 8520, Université Polytechnique Hauts de France, Université de Lille, CNRS, Valenciennes, France

Contents

10.1 Introduction

In recent years, deep learning methods have been used in many applications, including pattern recognition, computer vision, natural language processing, and speech recognition [1]. Due to the exponential

Artificial Neural Networks for Renewable Energy Systems and Real-World Applications.
DOI: https://doi.org/10.1016/B978-0-12-820793-2.00001-X

increase of data, deep learning techniques produce results comparable to and in some cases surpassing human expert performance, not only in the above domains but also in bioinformatics, drug design, medical image analysis, and others. However, the success of machine learning methods in general, and deep learning in particular, relies on the fact that they operate on entire datasets at once without accounting for the situation where the dataset dynamically grows over time [1]. For this reason, combining deep learning techniques with continual learning methods allows the resulting models to learn continuously, accumulate the knowledge learned in previous tasks, and use it to help future learning.

Incremental learning is a part of continual learning [2], where the objective is to train a model, previously built to perform an ensemble of tasks, to learn new tasks and allow it to obtain high performances on both new and old tasks. To do this, a simple solution consists of retraining the target model from scratch every time new data corresponding to new skills become available. This training is performed using old and new data jointly. As simple as this solution is, for many reasons it is not always feasible, due to the time spent in training the new model, in addition to the waste of power and computing to accomplish it. Another drawback to this practice remains the major difficulty of obtaining previously seen data each time the model needs to be trained. Apart from the fact of storing the data, and even with huge storage capacities, providing old data might be impossible in multiple situations due to privacy, such as for biomedical data.

Feature extraction and *fine-tuning* may also be helpful while a model attempts to learn new skills. There are special cases of domain adaptation (when old and new tasks are the same) or transfer learning (for different tasks) [3], which use knowledge acquired from one task to help learn another. The first model can be seen as a pretrained convolutional neural network (CNN), which is divided into two parts, one performs the feature extraction and the other is responsible for the classification task. In the context of transfer learning, on one hand, the *feature extraction* part of the previously trained network is maintained and the classification part only is retrained on different datasets [4] to help the model in learning new tasks. On the other hand, *fine tuning* adapts a network to new data or tasks, by adjusting the output layer of the original network, either by replacing it with classes corresponding to the new task, or by adding new classes to the existing ones [5]. For both *fine-tuning* and *feature extraction* the final model scores good performance on the new tasks and degrades drastically on the old ones. This issue, where a neural network forgets previously learned knowledge when adapted to a new task, is referred to as *catastrophic interference* or *forgetting* [5].

Besides *catastrophic forgetting*, many challenges face an incremental learning algorithm such as online model parameter adaptation caused by the fact that datasets are not available to start, but arrive over time as for biomedical data. *The stability—plasticity dilemma* is where learning new tasks requires plasticity for the integration of new knowledge, but also stability in order to prevent the forgetting of previous knowledge. *Concept drift* and *model benchmarking* also remain as challenging issues for incremental learning.

The existing approaches to continual learning that mitigate catastrophic forgetting are divided into three families *replay-based*, *regularization-based*, and *parameter isolation-based methods* [2].

10.1.1 Replay-based methods

Methods in this category mitigate forgetting by relying on storing previous experience explicitly or implicitly wherein the former raw samples are saved into the memory for *rehearsal*, whereas in the latter a generative model such as a generative adversarial network (GAN) [6] synthesizes them to perform *pseudo-rehearsal* [7]. The stored samples from previous tasks are replayed while learning a new task to alleviate forgetting. These samples/pseudosamples can be either used for rehearsal, approximating the joint training of previous and current tasks, or to constrain the optimization of the new task loss not to interfere with the previous tasks [2], such as for iCaRL (Incremental Classifier and Representation Learning) [8].

10.1.2 Regularization-based methods

When no storage of raw input is possible, mainly motivated by privacy reasons, as in the case of medical applications, regularization-based methods are memory efficient and propose an extra regularization term in the loss function to consolidate previous knowledge when learning on new data, such as for LwF [9], elastic weights consolidation (EWC) [10], and SupportNet [11].

10.1.3 Parameter isolation-based methods

When optimal performance is a priority and the model's capacity is not constrained on the architecture's size, parameter isolation-based methods perform by freezing the set of parameters learned after each previous task and growing new branches dedicated for new tasks. These methods are better suited for learning a long sequence of tasks, such as for Tree-CNN [12] and ACL_{seg} [13].

In this work, we discuss five incremental methods among the previously mentioned families. These methods are presented in two sections, where the first concerns the incremental learning of CNNs in general, and the second is about the same but in bioinformatics field.

10.2 Incremental learning of convolutional neural networks

According to Ref. [14], there are three continual learning scenarios: *Task-IL* for task-incremental learning, *Domain-IL* for domain-incremental learning, and *Class-IL* for class-incremental learning. The difference consists mainly of whether, at test time, information about the task identity is available, and if it is not whether the model is also required to explicitly identify the identity of the task it has to solve. Despite the existence of these three scenarios, the community has been most interested in *class-incremental learning*, and a large amount of methods have been proposed in this direction. In this chapter, we focus on the most popular.

10.2.1 iCaRL: incremental classifier and representation learning

Incremental Classifier and Representation Learning uses a neural network for feature extraction and performs classification based on a nearest-class-mean rule in that feature space, whereby the class means are calculated from the stored data. During the training of the feature extractor, iCaRL replays the stored data as well as the current task input with a special form of distillation [14]. To do this, iCaRL learns classifiers and a feature representation from a sample sets: X_1, X_2, ... where all samples of a set $X_y = x_1^y, \ldots, x_y^{ny}$ are of class $y \in \mathrm{N}$.

As soon as new data are available, iCaRL adjusts its internal knowledge *parameters* and *exemplar set*. Based on these new training examples, a memory size K, and current model parameters θ, the algorithm reduces the number of the exemplar set by $m = K/t$, where t is the class number that the iCaRL will be able to classify. The reduction is performed by keeping only the m first exemplars that are present in each exemplar-subset. Note that, for each class, the exemplar set is constructed around its feature mean exemplar μ, computed as follows:

$$\mu \leftarrow \frac{1}{n} \sum_{x \in X} \varphi(X)$$

where $\varphi : X \rightarrow R^d$ d is the *feature function* given by the current model, which is a CNN interpreted as a trainable feature extractor followed by a single classification layer with as many sigmoid output nodes as classes. The constructed set contains only data exemplars that *argmin* the distance between the current exemplar *feature-vector* and the *class-mean-exemplar-feature*.

For the classification task, although one can interpret the CNN model's outputs as probabilities, iCaRL uses the network only for representation learning. The classification task is performed applying a *nearest-mean-of-exemplars* strategy. The last relies on the *average-feature-vector* μ_y which represents a prototype of each class observed so far:

$$\mu \leftarrow \frac{1}{|P_y|} \sum_{p \in P_y} \varphi(P)$$

where P_y is the exemplar-set for y.

In order to predict an image x, its *feature-vector* is computed and the iCaRL assigns it the label class of the most similar prototype *feature-vector.*

$$y^* = argmin\|\varphi(\mu_y),\ \varphi(x)\|$$

Regarding the *representation-learning* update, iCaRL constructs an augmented training set consisting of the currently available training examples together with the stored exemplars. After that, the current network is evaluated for each example and the resulting network outputs of all previous classes are stored. The update is performed by minimizing a loss function that for each new image encourages the network to output the correct class indicator for new classes (classification loss), and for old classes, to reproduce the scores stored in the previous step (distillation loss).

For tests, the authors propose three class-incremental methods, as described below.

Fine-tuning: as in transfer learning [3], this learns a multiclass classifier for new incoming classes by fine-tuning the previously learned multiclass classification network.

Fixed-representation: aims to learn a multiclass classification network. This is achieved by freezing the feature-representation and the weights of the classification layer, the former after the first batch of classes has been processed and the latter after the corresponding classes have been processed. However, for subsequent batches of classes, only the weights vectors of new classes are trained.

LwF.MC refers to a multi-class classification using the *LwF* [9] algorithm which is discussed in the next section. The mentioned algorithm uses the distillation loss during learning, as iCaRL does, but without the need for an exemplar-set.

The above alternatives of class-incremental methods, *fine-tuning, fixed-representation*, and *LwF.MC*, were evaluated on the CIFAR-100 [15] dataset, and the ImageNet ILSVRC 2012 dataset, as described in Ref. [8]. The obtained results showed that iCaRL's predictions are distributed close to uniformly over all classes, whereas *LwF.MC* tends to predict classes from recent batches more frequently. The classifier with *fixed-representation* has a bias toward classes from the first batch, while the network trained by *fine-tuning* predicts exclusively class labels from the last batch as a direct consequence of catastrophic forgetting. Even with a complex and time-consuming algorithm for constructing an *exemplar-set* in one hand, and the need for the memory M management on the other, the iCaRL remains the benchmark algorithm for *replay-based* methods.

10.2.2 LwF: learning without forgetting

The main objective of the learning without forgetting [9] algorithm is to add for a new task n, new parameters θ_{new}, for a given CNN that have shared parameters θ_{shared} and task-specific parameters θ_{old}. The new CNN has to learn parameters that work well for both old and new tasks using images and labels from only the new task.

Divided into three steps, the algorithm starts by computing Y_{old}, the current CNN's response for new available images. Y_{old} corresponds to a set of probabilities of belonging to classes. In the second step, a copy of the current network is saved and the algorithm adds nodes to the CNN's output layer, for each new class. Weights corresponding to the added nodes θ_{new} are randomly initialized. After that, the θ_{shared} and θ_{old} weights are frozen and the network is trained on new images to convergence. Finally, the network jointly trains all weights (θ_{new}, θ_{shared} and θ_{old}) until convergence to minimize the loss for all tasks and regularization using the stochastic gradient descent as follows.

For *new tasks*: the loss encourages the predictions $\hat{y}_{new}$ to be consistent with y_{new}:

$$L_{new}\left(y_{new},\ \hat{y}_{new}\right) = -\ y_{new} * \log \hat{y}_{new}$$

where: $\hat{y}_{new}$ is the set of probabilities returned by the network, and y_{new} is the one-hot ground truth label vector. Both are considered for a signal image, but the total loss is averaged over all images in a batch while training.

For each *old task*: the goal is that the network outputs a probability close to the response obtained from the original network. For this purpose, knowledge distillation proposed in Ref. [16] is used by increasing the weights for smaller probabilities to encourage the output of one network to approximate the outputs of another.

For the *old tasks*, the starting point is a well-trained CNN which implies a large-scale dataset, such as the ILSVRC 2012 subset of ImageNet [17] which has 1000 object category classes with more than 1,000,000 training images. The Places2 [18] taster challenge in ILSVRC 2015 has 401 scene classes with more than 8,000,000 training images.

In contrast, for the *new tasks*, datasets with a moderate number of images are used, namely the *PASCAL VOC 2012* image classification [19] which contains 5717 images, the *Caltech-UCSD Birds-200—2011 fine-grained classification* [20] with 5994 images, the *Massachusetts Institute of Technology (MIT) indoor scene classification* [21] containing 5360 scenes and the handwritten characters dataset Modified National Institute of Standards and Technology database of handwritten digits (MNiST) [22].

The three common learning strategies for incremental learning are evaluated, namely: *feature extraction*, *fine-tuning*, and *joint-training*. In the first, θ_{shared} and θ_{old} are unchanged, and the outputs of one or more layers are used as features for the new task in training θ_{new}. In contrast, for *fine-tuning* θ_{shared} and θ_{new} are optimized for the new task, while θ_{old} is fixed. A *joint-training* strategy jointly optimizes all parameters θ_{shared}, θ_{old}, and θ_{new}.

Apart from the fact that LwF does not require access to the training data for previous tasks, it is faster to train and performs similarly to joint training. In the *single new task* scenario and for *new tasks*, Lwf outperforms fine-tuning and feature extraction except for ImageNet to MNiST since the handwritten characters are completely unrelated to ImageNet classes. However, the algorithm performs better than fine-tuning for *old tasks*.

On the other hand, and in the context of a *multiple new task* scenario, LwF outperforms *fine-tuning* on all tasks, outperforms *feature-extraction* for new tasks, and performs similarly overall to *joint-training* except on the old tasks in ImageNet to scenes.

10.2.3 Tree-CNN: tree convolutional neural networks

Inspired by hierarchical classifiers, Tree-CNN [12] is composed of multiple nodes connected in a tree-like manner. At the root node, a deep convolutional neural network (DCNN) ensures the first classification by assigning labels to input images. Each node in the tree contains a DCNN trained to classify images into one of its children nodes. Leaf nodes are associated with final classes. The authors start from a DCNN trained to recognize a certain number of objects that act as a root node. As soon as a new task T is available, the goal is to learn to identify images belonging to the M new classes present in the new task T. For each new class, about 10% of sample images are fed into the root node DCNN, and the K output results corresponding to the DCNN children number are used to build the O^{K*M*I} matrix, where I is the number of sample images per class.

The O^{K*M*I} matrix is used to compute the O^{K*M}, which is the average of the outputs over I images. After that, a softmax likelihood matrix is computed over that O_{avg} in order to generate a list S of M objects, where each object $S[i]$ has three attributes:

$S[i]$.*label*: contains the label of the new class;

$S[i].value = [v_1, v_2, v_3]$: corresponding to the top three average softmax (O_{avg}) output values for that new class in descending order, that is, $v_1 \geq v_2 \geq v_3$;

$S[i].nodes = [n_1, n_2, n_3]$: are the output nodes corresponding to the softmax outputs $v_1, v_2,$ and v_3.

In order to ensure that new classes with high likelihood values are added first to the Tree-CNN, the generated list S is ordered in decreasing value of the $S[i]$ *value* attribute, so that the first element's value allows the algorithm to take one of the following three actions:

Add the new class to an existing child node: if $v_1 \geq v_2$ by α a user-defined threshold indicating a strong resemblance with a particular child. The new class is added to the corresponding child node n_1 which needs to be retrained.

Merge two child nodes to form a new child node and add the new class to this node: if $v_1 - v_2 \leq \alpha$ and $v_2 - v_3 \geq \beta$, where α and β

are thresholds defined by the user indicating that the new class has a strong likelihood for more than one child node.

Add the new class as a new child node: if all the child nodes are full (or, if $\nu_1 - \nu_2 \leq \alpha$ and $\nu_2 - \nu_3 \leq \beta$), the former avoids Tree-CNN's expansion in depth, and the latter indicates that the new class does not have a likelihood value that satisfies the α and β user's thresholds.

When performing predictions, the DCNN present in the root node returns its final class label or the final class label of the DCNN of one of its children.

The Tree-CNN algorithm was evaluated on the CIFAR(100) [15] dataset and compared for its test accuracy against LwF and iCaRL on one hand, and for the training effort on the other. The last was evaluated among five different depths of backpropagation when retraining with the incremental data and the old data. The results showed that the most optimal trade-off between training effort and testing accuracy for the Tree-CNN was 66%–70%, while iCaRL and LwF gave 65% and 45% test accuracies, respectively.

10.3 Incremental learning of convolutional neural networks in bioinformatics

In recent years, despite the success achieved by CNNs in computer vision applied in bioinformatics, especially for computer-aided diagnosis (CAD), in most biological datasets, the amount of data is regularly growing and the number of classes is continuously increasing [11]. Retraining a CNN from scratch using old and new data is highly computationally costly and the extracted features are likely very different from those extracted by the model trained on the old data alone [11]. Building powerful models is no longer a difficult task, but updating their knowledge is. In the following we discuss two algorithms which learn CNNs incrementally on biomedical image datasets.

10.3.1 SupportNet: a novel incremental learning framework through deep learning and support data

SupportNet [11] is an incremental learning framework for biological data through deep learning and support data that combines the strength of deep learning and support vector machine (SVM), so that a model can review the essential information of the old data when learning the new information. This combination is formed based on three distinct parts, namely deep *feature-extraction*, *support-data-selector*, and *regularizers*.

Feature-extraction: As soon as the need for transfer learning is present, the feature extraction becomes of vital importance. For this purpose, the Support-Net framework uses the SENet [23] deep learning architecture, which considers both the spatial and the channel information to obtain a good-structured high-level representation of the original input data. The resulting feature vectors are then used to produce the support data candidates.

Support-data-selector: During training, and for each class, support data are maintained then fed into the model with new data for new classes. This practice allows the model to review the representations of previous classes when learning new information. The target support is obtained by training a SVM classifier on feature-vectors of an old dataset to detect the support vectors which are of crucial importance for the classification. The original data which correspond to the support vectors are considered (defined) as the support data candidates.

Since SVM is not as powerful as deep learning on a multiclass classification, especially when the number of classes is large, the support data selector is not used as the final classifier, despite its training being considered as an approximation of the last layer of the deep learning mode, this SVM is not used as a final classifier.

Consolidation regularizers: In order to overcome the fact that previous support vectors for the old data may no longer be support vectors for the new data leading to the invalidity of the support data on one hand, and the model overfitting due to the high expressability of the extracted features and the limited size of the support data on the other hand, two consolidation regularizers for learned knowledge are introduced, the *Feature-regularizers* which force the model to produce fixed representation for the old data over time, and the *EWC-regularizers* which consolidate significantly the weights and contribute to the loss function of the old class classification.

The SupportNet network is compared with retraining from scratch when new classes become available, *fine-tuning*, iCaRL [8], and EWC [10], in addition to a method which assigns the label of each test data sample randomly without using any model. Note that retraining from scratch with all data can be considered as the empirical *upper bound* of the performance of the incremental learning methods.

Comparisons are performed on three different datasets. First, was the *Enzyme Function Prediction* dataset [24] which predicts the function of enzymes from their sequences. The obtained results showed that, for this task, the SupportNet framework stays stable in term of accuracy classification when the number of classes increases, compared to iCaRL and EWC. About 90% accuracy was obtained when learning six classes incrementally, which is less than a 5% difference compared to the *all-data upper-bound* method.

The second dataset was the *Breast Cancer Histopathological database* (BreakHis) [25] containing microscopic images which are classified into eight classes (four for benign samples and four for malignant ones). The third one is the *2D HeLa Images* dataset [26], which contains the fluorescence microscope images of the 10 major subcellular structures in HeLa cells used for the breast tumor classification task. Meanwhile the second dataset was used for a subcellular structure classification task.

For these two classification tasks, similar conclusions are drawn: the *all-data* method performs the best and SupportNet is a close second. Although iCaRL was specifically designed for image classification in computer vision, SupportNet outperformed it on these two biological image datasets, even with such a small dataset size that often results in overfitting for deep learning methods. The authors demonstrated that iCaRL is overfitted to the real training data (new data plus the exemplars) but underfitted to all the training data (new data plus all the old data) compared to SupportNet, where the performance degradation from the real training data to all the training data reduces from 37% using iCaRL to 7% using SupportNet. This suggests that the data candidate selected by SupportNet captures accurately the distribution of all the training data compared to iCaRL.

10.3.2 Continual class incremental learning for computerized tomography (CT) thoracic segmentation

The adversarial continual learning (ACL) segmentation ALC_{seg} uses the ACL [7] approach initially developed for incremental classification problems on MNiST [22] and CIFAR [15], and adopts it to solve segmentation problems [13] where the objective is to learn a sequence of T tasks one task at time. The adaptation generates an architecture containing, in addition to the *discriminator* and the *private-module* proposed in Ref. [7], a modified *private-head* and a modified *shared-module* as follows:

Shared-module: Due to the size of medical data, and the need to segment small anatomical structures, an encoder-based module with Atrous Spatial Pyramid Pooling (ASPP) [27] is used that allows explicit control over the resolution of extracted features (from Ref. [7]'s *shared-module*) and capturing multicontext information at reduced computational cost.

Private-head: On one hand, *shared* and *private* features are multiplied and added instead of being concatenated as in Ref. [7]'s head module. After that, the obtained results are concatenated over the channel dimension and then passed to the *task-head*. On the other hand, and in order to solve the checkerboard artifacts and to improve the segmentation results, the *private* and *shared* embeddings are upsampled in two stages with a $4\times$ upsampling factor at each stage. The Convolution Transpose upsampling is replaced with subpixel convolutions. These approaches are used to design a compact task head in order to make the model scalable each time a new task head is added for a new added task.

The ALC seg ALC_{seg} is evaluated on the American Association of Physicists in Medicine (AAPM) thoracic autosegmentation challenge dataset AAPM [28] using a *Dice-Coefficient* [13] metric to report the segmentation quality, and a modified *knowledge-retention* [29] metric to fall within [0,1], measuring the model's ability to learn new information. The AAPM training set is composed of 30 scans, which are split into 5 subsets (one for each class), 6 validation and 24 testing CT scans. The segmentations concern 5 organs: spinal cord, right lung, left lung, heart, and esophagus.

Experimental results against the LwF [9], a naive *fine-tuning* as *lower-bound* and a *joint-training* in a multitask setting with all available data as *upper-bound* show that although LwF is able to retain the performance of the base task, it struggles to accommodate new information as more classes are added. In contrast, the *fine-tuning* approach learns well the new classes, at the expense of losing previous ones.

However, ALC_{seg} is able to combine both capabilities by retaining a consistent performance on all the previously learned tasks with a *knowledge-retention* of 0.96 for the final class incremental learning on five classes, while *fine-tuning* and LwF record 0.93 and 0.82, respectively. ALC_{seg} is able to reach near-ideal performance on subsequent tasks with a *Dice-Coefficient* of 0.80 versus 0.75 and 0.14 for LwF and *fine-tuning*, respectively.

Regarding the task order robustness, experimental results show that starting with a hard-to-segment task, esophagus, for example, has an effect on the base score as the model was not able to fully learn the base class. However, the model was able to maintain the performance on the same class till the end of the incremental training.

10.4 Discussion

In this work, we review incremental learning of CNNs in recent literature, where all the methods are unanimous on the fact that the *fine-tuning* approach learns new classes well, at the expense of losing previous ones. Such a phenomenon is commonly known as catastrophic inference or forgetting. Experimental results are performed using different test strategies on several datasets (in terms of size and domain) with many evaluation metrics (see Table 10.1), which emphasizes the benchmark challenge that faces continual learning in general and incremental learning in particular.

Apart from the CIFAR(100) [15] dataset used by both iCaRL [8] and Tree-CNN [12], where each uses a different metric for evaluation, the other works were divergent on the experimented datasets. Table 10.1 shows that, among the compared methods, LwF is the only algorithm that does not require access to the training data for previous tasks. The accuracy is the metric used for evaluation, and from experimental results we conclude that it is fast to train and performs similarly to *joint-training*. However, it struggles to accommodate new information as more classes are added, or if the features of the new task are completely unrelated to those of the old ones, such as for ImageNet to MNIST [22].

In contrast, the iCaRL [8] classification accuracy degrades with an increase in the number of new classes. This is explained by the fixed memory budget allocated to support all the previously learnt classes. Even if the *select-exemplar-set* procedure is computationally costly and the prediction is done by a *nearest-exemplar- classification* algorithm instead of the using a CNN, iCaRL [8] remains the benchmark algorithm for the incremental learning state of the art methods.

Both Tree-CNN [12] and ALC_{seg} [13] rely on deep learning for the classification part. In the former, when a new task is added into a branch, its CNN is trained from scratch, which requires old and new data. In the latter, even if the *knowledge-retention* is preserved, the task order affects the anatomy segmentation performance.

10.5 Conclusion

In this work, five incremental algorithms and frameworks for CNNs are reviewed, where two of them have been applied on

Table 10.1 Comparing CNNs' incremental algorithms.

Method	iCaRL	LwF	Tree-CNNs	SupportNet	ACL*seg*
Category	Replay-based	Regularization-based	Parameter-isolation-based	Regularization-based	Parameter-isolation-based
New task		performances	Good	Good	Good
Good	Good				
Old task		performances	Good	Good	Medium
Good	Good				
Storage requirement	Medium	No	High	Medium	No
Metrics	Accuracy	Accuracy	Accuracy and effort	Kappa score, accuracy, and macro F1-score	Dice coefficient [13] and knowledge retention [29]
Dataset	CIFAR (100) [15]	ILSVRC 2012, Places2, PAS-CAL VOC, 2012, Caltech-UCSD Birds-200–2011, MIT indoor scene	CIFAR (10 and 100) [15] and MNiST [22]	Enzyme function prediction dataset, 2D HeLa images, the HeLa image dataset, BreakHis dataset	AAPM [28]
Method	FE and nearest mean examplars classification	FE and DNNs	DCNNs	Feature extraction, SVM, feature regularization, and weight consolidation	ACL [7], ASPP [27], and subpixel convolutions

biomedical image classification. In the same context, the combination of CAD and incremental learning techniques, especially for computer vision, provides powerful systems that are, as with humans, able to learn over the time. Nonetheless, catastrophic forgetting remains a long-standing challenge, and we look forward to the day when CAD can learn to perform multiple tasks continually without them interfering with each other.

References

[1] H. Kashyap, H.A. Ahmed, N. Hoque, S. Roy, D.K. Bhattacharyya, Big data analytics in bioinformatics: architectures, techniques, tools and issues, Network Modeling Analysis in Health Informatics and Bioinformatics 5 (1) (2016) 28.

[2] M.D. Lange, R. Aljundi, M. Masana, S. Parisot, X. Jia, A. Leonardis, et al., Continual learning: a comparative study on how to defy forgetting in classification tasks, arXiv preprint arXiv:1909.08383, 2019.

[3] S.J. Pan, Q. Yang, A survey on transfer learning, IEEE Transactions on Knowledge and Data Engineering 22 (10) (2010) 1345, 1359.

[4] W. Mousser, S. Ouadfel, Deep feature extraction for pap-smear image classification: a comparative study, in: Proceedings of the 2019 5th International Conference on Computer and Technology Applications, 2019, pp. 6–10.

[5] K. Shmelkov, C. Schmid, K. Alahari, Incremental learning of object detectors without catastrophic forgetting, Proceedings of the IEEE International Conference on Computer Vision, IEEE, 2017, pp. 3400–3409.

[6] H. Shin, J.K. Lee, J. Kim, J. Kim, Continual learning with deep generative replay, in: Advances in Neural Information Processing Systems, 2017, pp. 2990–2999.

[7] S. Ebrahimi, F. Meier, R. Calandra, T. Darrell, M. Rohrbach, Adversarial continual learning, arXiv preprint arXiv:2003.09553, 2020.

[8] S.-A. Rebuffi, A. Kolesnikov, G. Sperl, C.H. Lampert, iCaRL: Incremental classifier and representation learning, Proceedings of the IEEE Conference on Computer Vision and Pattern Recognition, IEEE, 2017, pp. 2001–2010.

[9] Z. Li, D. Hoiem, Learning without forgetting, IEEE Transactions on Pattern Analysis and Machine Intelligence 40 (12) (2017) 2935–2947.

[10] F. Huszár, On quadratic penalties in elastic weight consolidation, arXiv preprint arXiv:1712.03847, 2017.

[11] Y. Li, Z. Li, L. Ding, Y. Hu, W. Chen, X. Gao, Supportnet: a novel incremental learning framework through deep learning and support data, 2018.

[12] D. Roy, P. Panda, K. Roy, Tree-cnn: a hierarchical deep convolutional neural network for incremental learning, Neural Networks 121 (2020) 148–160.

[13] A. Elskhawy, A. Lisowska, M. Keicher, J. Henry, P. Thomson, N. Navab, Continual class incremental learning for ct thoracic segmentation, arXiv preprint arXiv:2008.05557, 2020.

[14] G.M. van de Ven, A.S. Tolias, Three scenarios for continual learning, arXiv preprint arXiv:1904.07734, 2019.

[15] A. Krizhevsky, G. Hinton, et al., Learning multiple layers of features from tiny images, 2009.

[16] G. Hinton, O. Vinyals, J. Dean, Distilling the knowledge in a neural network, arXiv preprint arXiv:1503.02531, 2015.

[17] O. Russakovsky, J. Deng, H. Su, J. Krause, S. Satheesh, S. Ma, et al., Imagenet large scale visual recognition challenge, International Journal of Computer Vision 115 (3) (2015) 211–252.

[18] B. Zhou, A. Khosla, A. Lapedriza, A. Torralba, A. Oliva, Places2: a large-scale database for scene understanding, Arxiv preprint:[pending], 2015.
[19] M. Everingham, S.M. Ali Eslami, L.V. Gool, C.K.I. Williams, J. Winn, A. Zisserman, The pascal visual ob- ject classes challenge: a retrospective, International Journal of Computer Vision 111 (1) (2015) 98–136.
[20] C. Wah, S. Branson, P. Welinder, P. Perona, S. Belongie, The caltech-ucsd birds-200–2011 dataset, 2011.
[21] A. Quattoni, A. Torralba, Recognizing indoor scenes, 2009 IEEE Conference on Computer Vision and Pattern Recognition, IEEE, 2009, pp. 413–420.
[22] L. Deng, The MNIST database of handwritten digit images for machine learning research [best of the web], IEEE Signal Processing Magazine 29 (6) (2012) 141–142.
[23] J. Hu, L. Shen, G. Sun, Squeeze-and-excitation networks, Corr abs/1709.01507 (2017). arXiv preprint arXiv:1709.01507, 2017.
[24] Y. Li, S. Wang, R. Umarov, B. Xie, M. Fan, L. Li, et al., Deepre: sequence-based enzyme ec number prediction by deep learning, Bioinformatics (Oxford, England) 34 (5) (2018) 760–769.
[25] F.A. Spanhol, L.S. Oliveira, C. Petitjean, L. Heutte, A dataset for breast cancer histopathological image classification, IEEE Transactions on Biomedical Engineering 63 (7) (2015) 1455–1462.
[26] M.V. Boland, R.F. Murphy, A neural network classifier capa- ble of recognizing the patterns of all major subcellular structures in fluorescence microscope images of hela cells, Bioinformatics (Oxford, England) 17 (12) (2001) 1213–1223.
[27] L.-C. Chen, Y. Zhu, G. Papandreou, F. Schroff, H. Adam, Encoder-decoder with atrous separable convolution for semantic image segmentation, in: Proceedings of the European conference on computer vision (ECCV), 2018, pp. 801–818.
[28] J. Yang, G. Sharp, H. Veeraraghavan, W.van Elmpt, A. Dekker, T. Lustberg, et al., Data from lung ct segmentation challenge, The Cancer Imaging Archive (2017).
[29] R. Kemker, M. McClure, A. Abitino, T. Hayes, C. Kanan, Measuring catastrophic forgetting in neural networks, arXiv preprint arXiv:1708.02072, 2017.

CHAPTER ELEVEN

Hybrid Arabic classification techniques based on naïve Bayes algorithm for multidisciplinary applications

Mohammed Otair[1], Somaya Zacout[1], Laith Abualigah[1,2,3,4] and Mahmoud Omari[1]
[1]Faculty of Computer Sciences and Informatics, Amman Arab University, Amman, Jordan
[2]Hourani Center for Applied Scientific Research, Al-Ahliyya Amman University, Amman, Jordan
[3]Faculty of Information Technology, Middle East University, Amman, Jordan
[4]School of Computer Sciences, Universiti Sains Malaysia, Pulau Pinang, Malaysia

Contents

11.1 Introduction

Over the past few years, the evolution of the Internet has created a new world; the digital content world, where the virtual web is flooded

DOI: https://doi.org/10.1016/B978-0-12-820793-2.00004-5

with data and information at phenomenal speed, all these data need to be identified to be of any use [1].

This dilemma created a new need, the need to classify and categorize text data based on its content, search engines promoted the use of search keywords but inexperienced users of those search engines encountered many unexpected problems as to the number of hits on the keywords that might not even be related to their request [2]. Artificial neural networks (ANNs) have been employed in many areas of solar energy: for the design of a solar steam producing plant, for the calculation of a parabolic channel collector intercept agent and local concentration ratio, and the modeling and production forecast of solar water heating methods [3]. The literature has confirmed that ANNs can be a hopeful option to traditional methods in many engineering applications, power design systems, renewable energy systems, and manufacturing applications [4,5].

Therefore, the task of text classification or categorization (TC), which is the process of assigning one or more categories or classes to a document to help manage and sort documents for various reasons such as spam filtering, patient reports, healthcare organization, and news organization, among many other uses, has been found which provided better search results [6]. Text classification lies at the intersection between information retrieval (IR) and machine learning (ML).

Text classification goes through a number of steps to ensure the best result of the process; those steps can be simply summarized as data preprocessing, text classification, and evaluation and assessment. The Arabic language requires highly intelligent preprocessing techniques due to its complex morphology, its high inflectional, derivational nature, and the widespread amount of synonymous text in it [7,8]. Many researchers have applied different text categorization (TC) algorithms such as K-nearest neighbor (KNN), naïve Bayes algorithm (NB), support vector machines (SVM), decision trees (DTs), polynomial networks, ANNs, learning vector quantization (LVQ), maximum entropy, and many others, accompanied by variant preprocessing techniques including normalization, stemming, removal of stop words, light stemming, filtering all the non-Arabic text, and removal of punctuation marks, diacritics, and numbers [9].

Arabic text categorization/classification is of high importance due to the widespread use of Arabic text online, which means there has to be a way to categorize such text to ease the online searching and labeling of text [10]. Unfortunately, there has not been an optimal solution to this problem for a number of reasons, including the lack of a suitable taxonomy that confines browsing. The naïve Bayes algorithm has multiple issues which are: (1) incomplete training data; (2) continuous variables; and (3) attribute independence [11,12]. Therefore this chapter attempts to find the solutions to these issues.

This chapter implements and evaluates the effect of naïve Bayes-based hybrid techniques in the Arabic text classification technique pool to help expand this area of research and provide a reference to other researchers in the future for enhancement. The chapter aims to test the effect of hybrid Arabic classification techniques based on naïve Bayes; one of these hybrid techniques is named NBC 4.5 that is based on naïve Bayes and C4.5. The goal of this technique was to reduce the complexity of the NBTree technique and to improve the overall accuracy of the classification. This technique showed an average accuracy of 85.7% when tested on 36 datasets by its creator; therefore this thesis attempts to evaluate the impact of naïve Bayes hybrid techniques such as this one on the categorization of Arabic text. Moreover, the proposed algorithms combined the naïve Bayes algorithm with different algorithms to find the best algorithm to overcome the main challenges of the naïve Bayes algorithm. The task of text classification for Arabic text within digital content is discussed; unfortunately, work in the classification of Arabic text remains limited, even though the Arabic language is considered the fifth most spoken language globally, with more than 4 million speakers around the world. The results showed that the voting NBSVM obtained better results than other algorithms.

The remainder of this chapter is organized as follows. Section 11.2 presents more related works in that domain; Section 11.3 presents the proposed method and its main procedures.

11.2 Related works

This chapter is dedicated to explaining the concepts of text classification and Arabic text classification in detail, and reviews related research and studies concerning this topic.

11.2.1 Data mining

Data mining in simple terms means mining or extracting information and knowledge from a vast variety of data sources [13]. This task holds another name, "knowledge mining from data," which indicates the main task of finding knowledge and information from the accumulated data in order to solve problems, or uncover hidden opportunities.

Data mining tasks are of two types, descriptive and predictive. Descriptive data mining results in the way a human can analyze and understand their associations, which allows the user to describe general properties of the data. On the other hand, predictive data mining predicts some response of interest by adding a specific response variable in data. It performs inference on the data available to make predictions. Some of the most commonly used predictive data mining methods are decision rules, DTs, logistic regression, SVM, ANNs, Bayesian networks and k-nearest (kNN) algorithms [14].

11.2.2 Text classification

Text classification (TC) is considered as one of the most important fields nowadays, with text classification/categorization being concerned with allocating text into a specific class or category, based on its content [15].

Manual text classification is considered almost impossible currently, and that is because the amount of publicly and digitally shared text, such as social media posts, blogs, articles, and many more, is massive; therefore the need for an automatic way to classifying all this content is of high importance. This field lies in the intersect between the sciences of IR and ML. TC is divided into two categories: supervised text classification and unsupervised text classification [16]. Text classification consists of several small problems; assigning weights to terms, indexing, reducing dimensionality, clustering of documents, determining thresholds, and choosing the classifier [17].

11.2.2.1 Information retrieval using machine learning

IR systems are systems that retrieve documents from huge text-based databases; these documents are searched for when the user submits the correct description for what they are looking for. Users constantly improve the descriptions they submit to accurately find what they need; retrieval systems can employ user feedback on chosen documents to show the accurateness of the description. The way the user submits or chooses the keywords of the description, and the way they are used to search the document database, are based on the indexing method used in collecting the documents. IR has four phases: indexing, querying, comparison, and feedback [18].

IR systems are abstractly divided into three collaborative subsystems; administrator, indexation, and user subsystems. Each subsystem has a specific task that will eventually retrieve the document needed. Administrator subsystem makes sure that documents are submitted into the system, then decide the base of the documents based on the set of documents available; the representation and storage method of the documents are then set by the document base manager. The indexation subsystem handles the formation of the index and the creation of the representation of the query that allows the comparison between the query and the index of the document. The user subsystem has the task of processing the query submitted by the user and then finding relevant documents. The whole IR system simply works as follows [18]:

1. The user submits a query that gets processed by the user subsystem and then gets a keyword assigned to it.
2. After that the indexation subsystem takes over the task by indexing that query and compares it to the index of the documents.
3. Then the administrator subsystem retrieves documents that are related to the query and sends them to the user.
4. In order to enhance this process the user submits a query called feedback by marking the most relevant documents to their query and sending it back for another comparison to receive more relevant documents. These steps are summarized in Fig. 11.1.

The processes of IR employed ML methods starting from the end of the 1980s. It started to automate some of the processes such as user modeling, finding information from queries and document classification due to the complexity of natural language, which helped dramatically in reducing the amount of work the experts needed to do and also limited the amount of errors and inconsistencies [20].

IR systems require the extraction of an adequate amount of information from the documents in order to correctly work and categorize the documents, therefore the system must build models that allow it to properly create the categories. Such models try to effectively capture the meaning of the documents to determine where to categorize them, an example of such models is the term vector models that try to categorize based on a group of terms or words [21].

Documents that have a given group of terms each match to a specific point in n-dimensional space, where n is the amount of terms. A term vector could be a Boolean vector indicating the set of words in a document, or a numeric vector that indicates how frequent a term appeared in

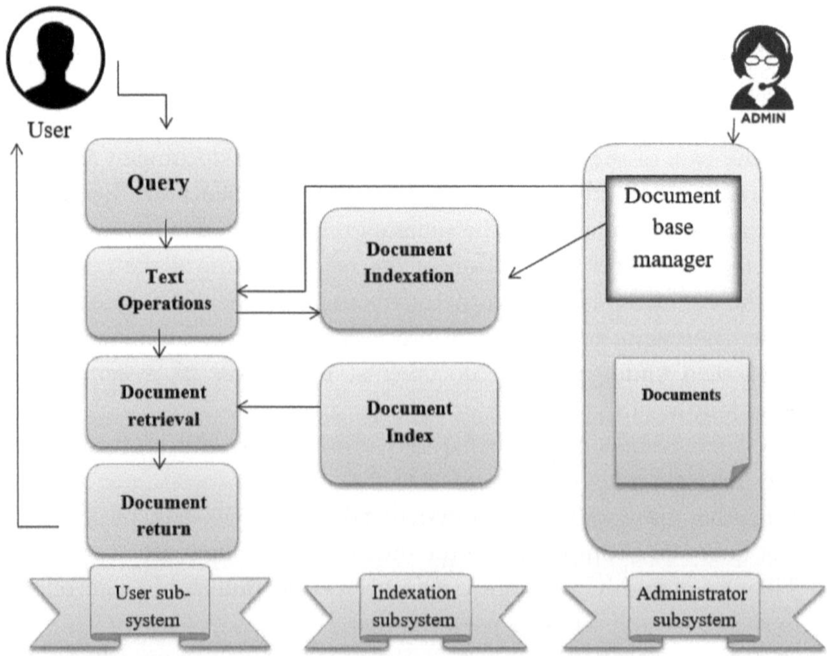

Figure 11.1 Information retrieval systems process [19].

a document. Following the model-building process comes the feature extraction phase where features are mined to aid in a better classification process. This phase is considered to be one of the most important phases in the learning process [22,23]. Feature selection is tricky due to the extreme complexity of the natural language and the difficulty in determining the meaning of a word within a document [24,25].

11.2.2.2 Machine learning for text classification

Text classification has become a very important task in recent years due to the digitization of almost everything. Therefore the need for text classification became supreme, and conducting this task manually is strenuous, consuming both time and human effort [26]. Because of that, ML algorithms have been successfully employed to reduce that effort, and make it a less time-consuming task. ML created a whole new concept in text classification, it was not strictly limited to classifying text into categories for digital libraries and online news chapters [27].

The new trend steered this—in many ways crucial—task toward, for example, what is commonly termed as sentiment analysis, which now is widely used in keeping nations safe as helps detect red flagged terms that might indicate any security issues or threats. This is simply a small portion of what text classification can do [28]. In order for ML techniques to properly function in an efficient and effective manner, some tasks must be applied prior to handing the data set in concern to the classifier; basically such tasks are referred to as preprocessing techniques [29].

The reason why such tasks are conducted is that the classifier cannot possibly perform when features are high in number, as this will cause the classifier to consider features of the text that have no significant meaning or value, consuming a huge amount of time which defeats the purpose of the text classification process. ML algorithms vary in terms of the way they operate and their capabilities, among many other variables. To name a few of these classifiers, support victor machine has been widely used lately and has shown promising results in terms of precision, unfortunately, it is not very promising in terms of recall, but remains a very efficient technique.

Naïve Bayes classifiers are used fairly often because they are well known for their effectiveness and simplicity, these classifiers have faced a couple of setbacks due to their inability to model text well, but many researchers have proved that such setbacks can be easily avoided using simple fixes [29]. Other classifiers have been used in a variety of ways, with different settings and variations in order to yield the best results. The recent trend has been to create a mix of classifiers called a hybrid, which means combining two or more classifiers working concurrently with the aim of rectifying each other's short-comings.

Text classification difficulty differs based on the amount of classes—the more classes there are, the greater level of difficulty it becomes This leads to the need for a larger training set, in some cases some of the classes will be a little harder to classify, for a couple of reasons, but mostly because there is a lack of positive training examples for that particular class, and few predictive features for it [30]. In the following list are some observed challenges that face TC using machine leaning [31]:

1. Compound semantics of text must be obtained by the feature vector.
2. All features attained from the frequency of words have to be free of noise.
3. Machine learning model performance is at risk of inferior performance if the training examples are of low quality.
4. In some cases text classification goes toward being subjective when some of the classes are unknown.

11.2.3 Previous studies

Many researchers over the years worked on the text classification task, where they used a massive variety of classifiers and worked on comparing their performance and enhancing it by changing and altering their components and techniques in order to correct some short-coming in them. In this section an overview of such work is presented, taking into consideration Arabic text classification specifically [32].

As mentioned in the previous section, text classification has become a new world need due to the large amount of digitized text [33]. It is worth mentioning that text classification is a crucial part of text mining. Text mining basically has two aims, first, it aims at facilitating the process of finding information within textual data, second, it manages IR and text classification operations alongside managing the task of summarizing natural language [34].

It is agreed upon that the performance of text classifiers is affected by many elements such as the amount of data fed into the classifier, and the quality, relevance, and redundancy of the data, all of this will reflect on the time and cost of the classification process. When it comes to text classification, researchers focus mostly on four main classifiers: naïve Bayes, KNN, DTs, and SVM, however, it is worth noting that work is not limited to these classifiers, only more focused on them [35].

Abualigah et al. [36] led full trials to compare numerous term weighting methods with SVM and suggested a new term, weighting method, with the name *tf.rf* to enhance the term's selective power. They found the newly suggested *tf.rf* scheme to be considerably superior to other term weighting methods. Tan [37] suggested a novel fine-tuning method for TC, created two advanced algorithms that bear the names refined centroid classifier and refined naïve Bayes classifier, and concluded that the advanced classifiers accomplished substantial performance enhancement over the original SVM classifiers. Chen [38] suggested feature selection for naïve Bayes text classification, recommended two feature assessment metrics called multiclass odds ratio and class discriminating measure for the naïve Bayes classifier, and concluded that these results showed better results than other approaches.

With the massive and growing availability of online Arabic documents, the urgent need for classifying algorithms is rising. Classification aims to map text documents in a specific category or more, based on keywords and content. With that said, the fact that the work in this field is unfortunately fairly limited, many techniques have been explored and applied by various researchers, such as the use of SVMs KNN, neural networks (NNs), naïve Bayes (NB), and DTs. The basic and most common method of Arabic text classification was to first preprocess the data then apply the proposed classifying algorithms, although some researchers chose not to clean and preprocess their data, such as in Ref. [39].

Other researchers applied different and widely varying preprocessing methods along with the most frequently and commonly used methods, such as stop words removal and stemming. Many researchers improved the known techniques to best fit the complex nature of the Arabic language such as the stop word remover based on a finite state machine (FSM) by Al-Shalabi et al. [40], also the P stemmer which is a modified version to one of the Larkey's light stemmer algorithms that works by removing the prefix and keeping the suffix.

Previous researchers such as Thabtah et al. [41] suggested a KNN algorithm using two methods: (1) term weighting and (2) different variations of the vector space models approach. They used a different term weighting IDF, *TF.IDF*, *WIDF*, *ITF*, and log (*1* + *TF*) in a mutual pattern with VSM such as Dice, Cosine, and Jaccard. They concluded that Dice-based *TF.IDF* and Jaccard-based *TF.IDF* exceeded the other methods with recall, precision, and f-measure, macroaverage cosine (87.6), Dice (87.7), and Jaccard (87.7). They applied this algorithm to a dataset of a news chapter website without preprocessing.

Research conducted by Harrag et al. [42] aimed to build an Arabic corpus and test the ANN model. Because of the great number of unique vocabulary in the collected set, they used the singular value decomposition (SVD) method to identify the most suitable features for the classification. The results confirmed that the ANN paradigm using SVD achieves 88.33%, which exceeds the performance of basic ANN which outputs 85.75% on Arabic document classification This thesis was conducted on an Islamic dataset consisting of 453 documents with normalization, removed punctuation marks, precluding stop words, diacritics, and nonletters as preprocessing steps.

Belkebir and Guessoum [43] suggested a technique to enhance Arabic text classification based on three different methods: ANNs, supported vector machines (SVMs), and a hybrid approach (BSO-CHI-SVM). They demonstrated the approach application and assessment using two kinds of representations: light stemming and root-based stemming. The Arabic dataset used in their research is called Open

Source Arabic Corpora (OSAC), which consists of 1000 documents. They applied the following preprocessing steps: removal of Latin alphabet, applied exclude stop words, isolated letters, diacritics and elongation, remove the digits and punctuation marks, and normalization of "HAMZA." They concluded that the approaches based on the representation via a light stemmer slightly topped those based on a root-based stemmer. In addition, the approaches based on SVMs performed much better than those based on NNs. Especially, the hybrid BSO-CHI-SVM approach proved to be the most efficient one with regard to F1, precision measures, and recall.

For the evaluation methods according to Belkebir and Guessoum [43], precision, recall, and F-measure are the most repeatedly used evaluation measures. The fraction of recovered documents that is relevant to the query is called precision. Recall is the segment of the documents which is accurately related to the inquiry that is absolutely recalled. Strictly speaking, it is concerned with how complete the results are. F-measure is roughly the average of precision and recall (Table 11.1).

In this chapter, the effect of the naïve Bayes algorithm when merged with other algorithms is evaluated and measured to show its performance when merged with two different methods. The first group of algorithms (SVM, ANN, and J48) are each merged with the naïve Bayes algorithms using the Waikato Environment for Knowledge Analysis (WEKA) voting method to create hybrid algorithms. The second algorithm is the new NBJ48, which has been created by Muqasqas [45]. The proposed methods are explained in detail in Section 11.3.

11.3 The proposed method

In this section, a full discussion of the work method is presented and laid out in detail. Fig. 11.2 summarizes the steps before reading the details for easier comprehension of the process.

11.3.1 Collecting information

When working on the text classification problems one must consider reviewing and collecting as much knowledge and information on the matter. An earlier intensive study review has been conducted in order to

Table 11.1 Summary of previous studies.

References	Research name	Year	Idea	Results
[37]	An improved centroid classifier for text categorization	2008	Suggested a novel fine-tuning method for text categorization	They concluded that the advanced classifiers accomplished substantial performance enhancement over the original SVM classifiers
[38]	Feature selection for text classification with naïve Bayes	2009	Suggested feature selection for naïve Bayes text classification	Results showed better results than other used approaches
[41]	Naïve Bayesian based on chi square to categorize Arabic data	2009	Suggested KNN algorithm using two methods: (1) term weighting and (2) different variations of vector space models approach	Dice-based TF.IDF and Jaccard-based TF.IDF exceeded the other methods with recall, precision, and f-measure
[44]	Neural network for Arabic text classification	2009	Proposed using ANN to enhance Arabic text classification	Results confirm that ANN paradigm using SVD achieves 88.33%, which exceeds the performance of basic ANN which outputs 85.75% on Arabic document classification
[43]	A hybrid BSO-Chi2-SVM approach To Arabic text categorization	2013	Suggested a technique to enhance Arabic text classification based on three different methods	The approaches based on the representation via a light stemmer slightly outperformed those based on a root-based stemmer. In addition, the approaches based on SVMs performed much better than those based on neural networks

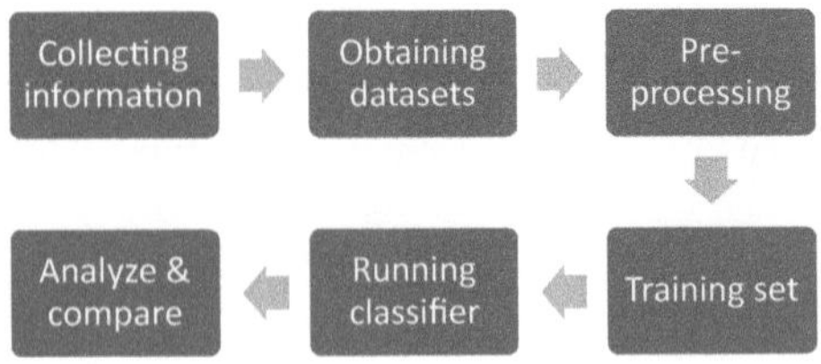

Figure 11.2 Thesis methodology phases.

gather as much needed knowledge as possible and to work on the right track for finding useful results. Many researchers worked on this paradigm and came up with various opinions and results, and tried to manipulate the settings of the classifiers and other variables in order to rectify or enhance the performance of the classifiers for a certain case.

The other part of the information collected was specifically concerned with the text classification problem within the Arabic language, which as mentioned before is one of the top five most spoken languages globally, being spoken in almost 25 countries with a little over 200 million people using it, giving it a high level of importance [46]. The Arabic language is a highly complex language that is made up of three types of words (verbs, nouns, and particles). It involves rich morphology and a compound orthography, with 28–29 letters it contains an almost endless supply of used words with all kinds of structures and meanings, that are extracted from a root set of 10,000 [47].

11.3.2 Obtaining datasets

In order to compare the work done here with other previous work, three datasets were chosen to implement the work on; these can be found in Ref. [48].

11.3.2.1 Open Source Arabic Corpus

Open Source Arabic Corpus (OSAc): this dataset has been gathered from more than one newschapter and news source, and contains 22,429 text documents, with 10 categories (History, Sports, Economics, Education & Family, Health, Religious and Fatwas, Law, Astronomy, Stories, and Cooking Recipes). The corpus contains about 18,183,511 (18 M) words and 449,600 distinct keywords after the removal of stop words [48].

11.3.2.2 BBC Arabic News

BBC Arabic corpus gathered from the BBC Arabic website has 4763 text documents, distributed among seven categories (International Press, Middle East News, World News, Sports, Business & Economy, Science & Technology, and Art & Culture). The corpus has 1,860,786 words and 106,733 distinct keywords after the removal of stop words.

11.3.2.3 CNN Arabic News

This corpus was collected by a researcher from the CNN Arabic website and chosen in this thesis with the aim of a more efficient comparison between results of the thesis in hand and previous work. It consists of 5070 documents distributed among six categories (business, entertainment, Middle East news, science and technology, sports, and world news). The corpus has 2,241,348 words and 144,460 distinct keywords after removing stop words.

11.3.3 Preprocessing

Preprocessing is one of the key phases of text classification. It refers to the process of converting text to data mining ready structure of that text, where it separates the text into individual words and presents each document as a feature vector.

11.3.3.1 Tokenization

The task of slicing up data into pieces is called tokenization. Tokens are the basic components for the dataset at the initial stage. All processes require tokens for text analysis and processing, for that reason, tokenization is a very important step to make data more reliable and easy to analyze [22,49,50].

11.3.3.2 P stemmer

A new Arabic light stemmer, a modified version of one of the Larkey's light stemmer algorithms, focuses on a collection of Arabic news articles. The collection classifies the article category using binary and multiclass methods and creates a new category called Taxonomy. The P stemmer works by removing the prefix and keeps the suffix, which gives much better results than Larkey's algorithms, especially by using a binary method and SVM. For example, (المكتبات) will be (مكتبات); in light10 stemmer it will be (مكتب) which is a different mean.

11.3.3.3 Term frequency—inverse document frequency

This term reflects the significance of a token in the corpus or dataset. Term frequency—inverse document frequency (TF-IDF)'s importance grows with the growth in frequency of a specific word in a document. To be able to control the generalization of more common words, the term frequency is balanced by the frequency of words in the corpus. Term frequency is the amount of times a particular term is found in the text. Inverse document frequency measures the existence of any word in all documents [51].

The confusion matrix or contingency table is a helpful parameter in measuring the performance of supervised data-mining algorithms. It is used in a supervised data-mining algorithm to help in evaluating the algorithm performance. From the classification point of view, terms such as "false positive (FP)," "true positive (TP)," "false negative (FP)," and "true negative (TN)," are used to compare labels of classes in this matrix. True positive indicates the amount of reviews those are positive and classified as positive by the classifier, whereas false positive indicates positive analyses, where the classifier does not classify it as positive. Also, true negative denotes the reviews which are negative, and classified as negative by the classifier, while false negatives are negative analyses which the classifier does not classify as negative. Machine leaning and data-mining technique performances are evaluated using parameters like precision, recall, f-measure, and accuracy [51].

11.3.4 The classification process

After obtaining all the information about the problem in hand and collecting the datasets that will be used and applying the preprocessing steps described earlier, the processed datasets are fed into the WEKA data-mining tool and the proposed classifiers are applied.

11.3.4.1 WEKA data mining tool

This tool (WEKA) was developed in the University of Waikato of New Zealand in order to create a more unified working platform with a very user-friendly interface for researchers and learners, to test their hypotheses and ideas regarding the use of ML algorithms and to provide easy access to preprocessing tools. This tool was developed using Java and is available for public use, and researchers can easily add to or modify it [52].

11.3.4.2 Classifying algorithms

Algorithms applied using the WEKA tool in order to test for this thesis are J48, SVM, and ANN, combined with the naïve Bayes (NB) algorithm using majority voting which is a two-step method. The first step uses subsets of the original data to produce a series of averagely performing models and the second "boosts" their performance by merging them together using a specific cost function, it is important to note that the tests have been conducted on all three datasets with the mentioned algorithm separately and combined with the naïve Bayes algorithm in order to show the effect of the naïve Bayes on the result.

11.3.4.2.1 Naïve Bayes algorithm

This classifier has been studied in detail since the 1950s, and is known for its extreme competitiveness in the area of text classification. In the area of ML, naïve Bayes classifiers are known as a group of simple probabilistic classifiers that are based on applying Bayes' theorem with strong (naïve) distinctive assumptions between the features [53].

Naïve Bayes algorithms have many advantages, including being highly recommended to use when working with numeric and textual data, and it is agreed upon that they are fairly easy to implement and use. Also, they have some weaknesses such as the real-world data violate the conditional independence assumption and the algorithm performs poorly when the features are highly correlated. The naïve Bayes algorithm has multiple issues including: (1) incomplete training data, (2) continuous variables, and (3) attribute independence. Therefore this study attempts to find solutions to those issues.

11.3.4.2.2 New NBJ48

This algorithm is a hybrid algorithm that is based on naïve Bayes and C4.5. The goal of this technique was to reduce the complexity of the NBTree technique and to improve the overall accuracy of the classification. It passes through four phases with multiple steps in these phases:

First phase: Building the naïve Bayes and C 4.5, after that the accuracy of the techniques is calculated using the 10-fold cross-validation.

Second phase: In this phase a comparison between the accuracies of both algorithms is made which is considered the main phase of this technique because this comparison determines if the C4.5 should stop splitting and set the naïve Bayes, or if the accuracy of C4.5 is higher and should continue splitting until there are 50 instances in each node of the tree, and so on.

Third phase: Splitting the dataset where it splits the nonleaf dataset from phase 2 till the end of the decision tree.

Fourth phase: Building the hybrid algorithm, rule-based pruning has been used to remove redundant and irrelevant nodes or subtrees, which improves accuracy and solves the outfitting issue.

11.3.4.2.3 Support vector machine

SVM is mainly a classifier method that performs classification tasks by building hyperplanes in a multidimensional space that splits cases of different class labels. SVM works with both regression and classification tasks and is able to handle multiple continuous and categorical variables. It works by discovering the linear separating hyperplane which maximizes the margin [i.e., the optimal separating hyperplane (OSH)]. SVM captures inherent characteristics better that the ANN algorithm, and minimizes the upper limit on the generalization error in a better way than the empirical risk minimization (ERM) principle, but SVMs need a lot if tuning for their parameters [54].

11.3.4.2.4 Artificial neural networks

ANNs or connectionist systems are computing systems that were inspired by the neural system in animal brains. Those systems learn in a progressive manner that improves their performance by using examples to learn. ANN has wide use in the real world such as medicine and biology. It can give a prediction and approximate to the results from data given, and always gives good results that are very close to the real results.

Usually, neurons are structured in layers. Each layer performs a different type of transformation on the input it receives. The path that signals travel from in these systems is from the first input all the way through to the last output layer, possibly after crossing the layers multiple times. Those systems begin with the goal of simulating the function of the human brain in solving the problems, but after a while the attention was more on specific sets of mental abilities, which led them away from biology, such as the concept of backpropagation, which indicates the ability of passing information in the opposite direction and then altering the network to reflect this information. ANN has many advantages such as producing excellent results in complex domains, and it is very appropriate for both discrete and continuous data with very fast testing capabilities. It also has some disadvantages, such as it trains in a very slow manner and the results obtained are difficult to interpret with the fact that ERM may lead to overfitting because it tries to make the algorithm reduce the error rate.

11.3.4.2.5 J48

J48 refers to the JAVA implementation of C4.5 algorithm and an extension to ID3 algorithm that were both created by Ross Quinlan. It is an algorithm used to build a decision tree and is considered one of the most used in practice [54]. It has many improvements over the old ID3 algorithm such as:

1. Managing attributes with variant costs;
2. Managing training data with missing values;
3. Managing discrete and continuous attributes.

The above-mentioned algorithms (SVM, ANN, and J48) were each combined with the naïve Bayes algorithm using the voting method in the WEKA tool, creating new hybrid classifying techniques names (Vote NBSVM, Vote NBANN, and Vote NBJ48, respectively). All the above steps are summarized in Fig. 11.3.

11.3.5 Evaluation

For the evaluation methods according to Ref. [55], precision, recall, and f-measure are the most repeatedly used evaluation measures.

1. Precision: This measures the accuracy of the classifier result as given in Eq. (11.1). It is the relation of the number of instances correctly categorized as positive to the total number of positively classified instances.

$$Precision = \frac{TP}{TP + FP} \tag{11.1}$$

2. Recall: This measures the wholeness of the classifier result as given in Eq. (11.2). It is the relation of the total number of definitely categorized instances to total instances which are truly positive.

$$Precisio = \frac{TP}{TP + FN} \tag{11.2}$$

3. f-Measure: This is the consistent mean of precision and recall as given in Eq. (11.3). It is required to improve the system toward either precision or recall, which has more of an effect on last result.

$$f - Measure = \frac{2 * Precision * Recall}{Precision + Recall} \tag{11.3}$$

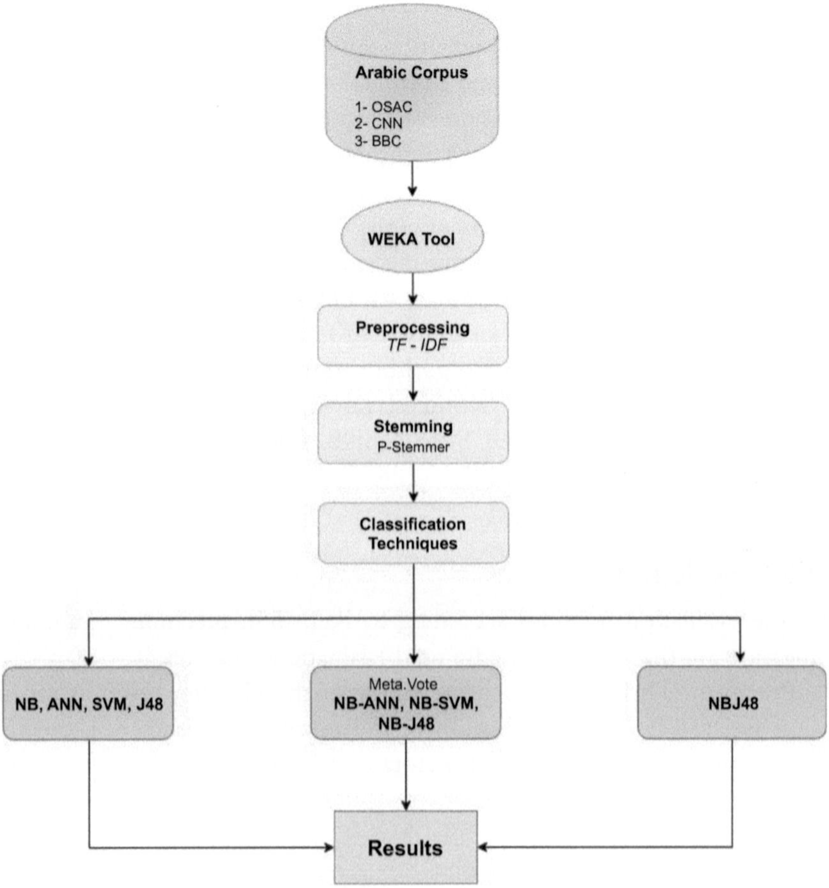

Figure 11.3 Methodology work flow using the Weka tool.

11.4 Results and discussion

Three datasets of different sizes were used in this thesis (OSAc, BBC, and CNN), with a total of 32,262 documents, using the WEKA tool after preprocessing the data the process of TC was conducted.

First, the performances of the four main algorithms were measured in order to be able to compare and measure the effect the naïve Bayes algorithm has on the performance of the TC process combined with other algorithms. Using the WEKA tool, naïve Bayes algorithm was merged with three algorithms (SVM, ANN, and J48) using the vote method, and a new programed algorithm that merged naïve Bayes algorithm with J48 in a different method as explained earlier.

11.4.1 Results

The results show that the size of the dataset has no effect on the compared performance of the algorithms, therefore the average for all the dataset results is calculated to show the actual and important information deduced from the experiment. In Fig. 11.4, the average time spent on building the model for each algorithm is shown. As is clearly shown in the figure, the ANN algorithm takes the longest time to build the model due to its complex nature, followed by the Vote NBANN as a natural result of the time ANN needs on its own but with a remarkable drop in time when combined with the NB algorithm. After that the New NBJ48 follows it with 1200 seconds needed for building the model.

In Fig. 11.5 we can see the results of the performances of the algorithms across all three datasets in terms of accuracy. The new NBJ48 shows phenomenal results, with an astonishing 98% accuracy it tops the next best algorithm, the SVM, by 3%, it is worth pointing out that the New NBJ48 algorithm has never been used before on Arabic text but it clearly shows extremely promising results.

Fig. 11.6 shows the average f-measure; the results here correlate with the previously displayed results following the same pattern where the new NBJ48 surpasses the next best algorithm by a little over 0.3.

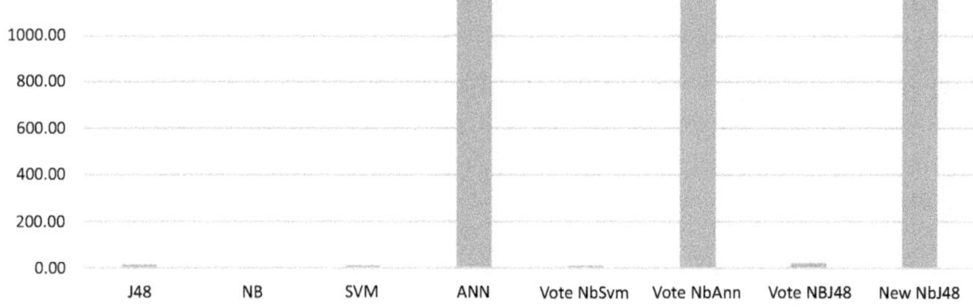

Figure 11.4 Average time spent for building models.

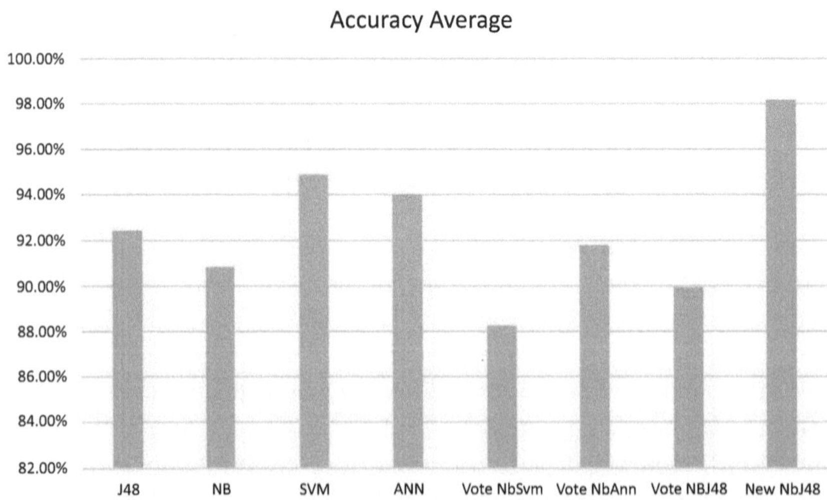

Figure 11.5 Average accuracy.

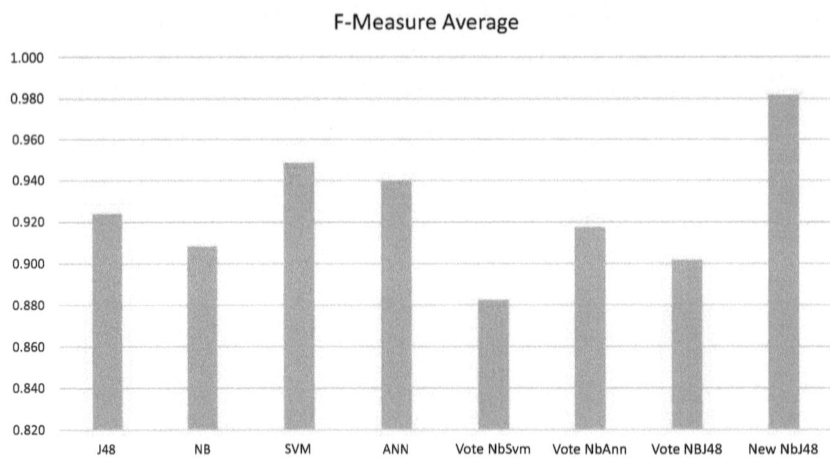

Figure 11.6 Average f-measure.

For the average recall and average precision in Figs. 11.7 and 11.8, respectively, we can deduce that the new NBJ48 tops the next best algorithm, SVM, by about 2.6% for recall and 0.63% for precision.

The following figures show the performance across all three datasets compared to each other. Fig. 11.9 shows the accuracy in all three datasets, where it can be seen that there is some kind of pattern across all algorithms except with the naïve Bayes algorithm in the OSAc dataset, where

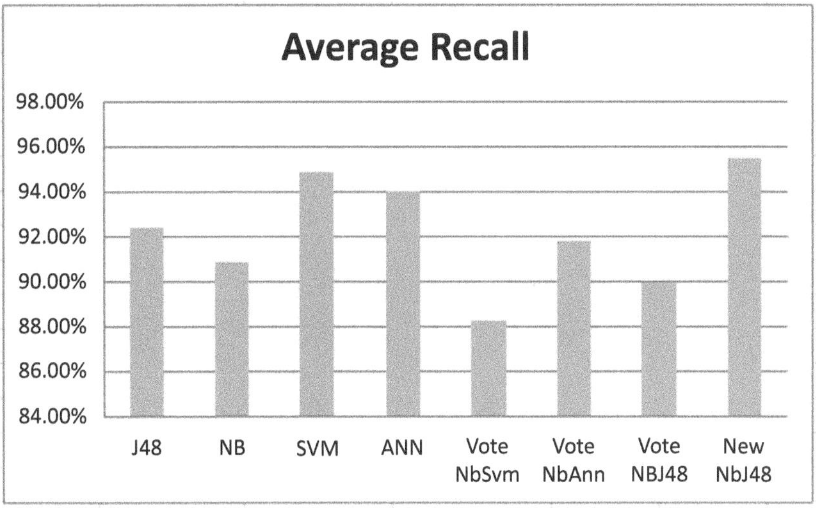

Figure 11.7 Average recall.

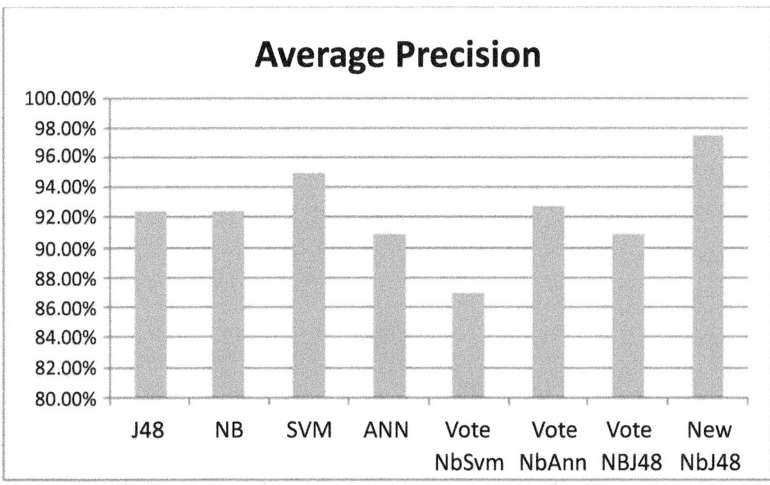

Figure 11.8 Average precision.

the accuracy drops instead of rising in comparison with the accuracy of the J48 algorithm accuracy in the BBC and CNN datasets.

In terms of f-measure, precision, and recall across the three datasets, Figs. 11.10–11.12 show correlating results with the previous figure where the performance also drops with the naïve Bayes algorithm in the OSAc

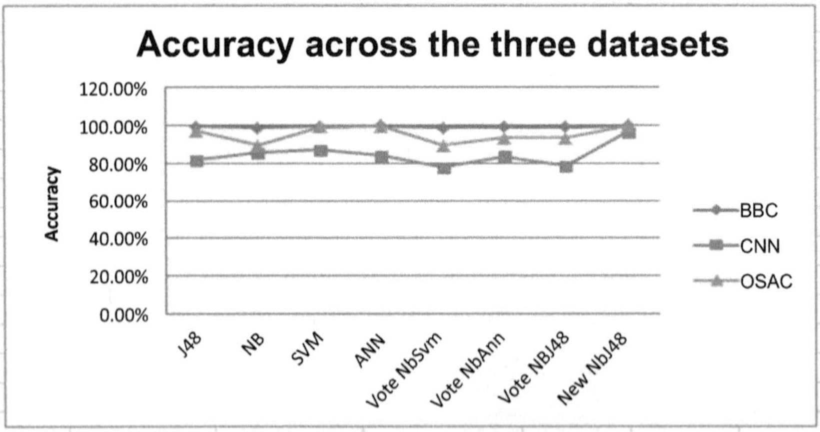

Figure 11.9 Accuracy in the three datasets.

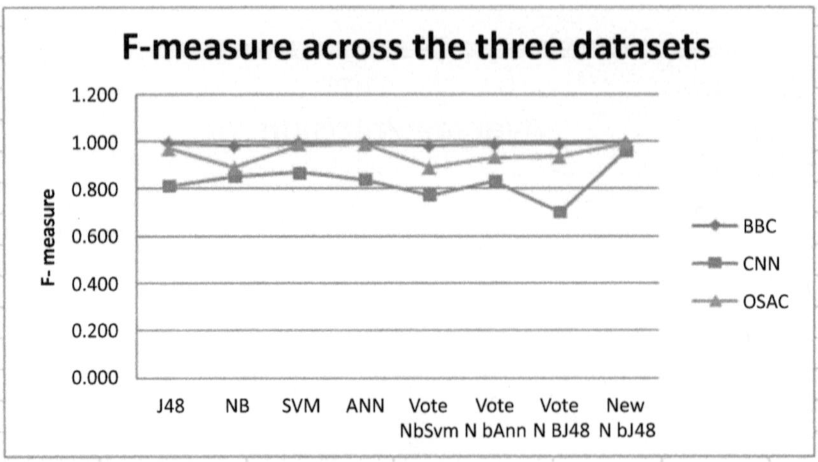

Figure 11.10 f-Measure across the three datasets.

dataset, which could be because the naïve Bayes algorithm is well known for its ability to handle large datasets.

All the above results show that the new NBJ48 algorithm, with its new technique, has a huge effect on the Arabic text classification, but unfortunately the naïve Bayes algorithm, when combined with other algorithms using the traditional vote technique available on the WEKA tool, degrades their performance. Although it is usually used because of its simplicity, its effect takes a toll on the performance of other algorithms.

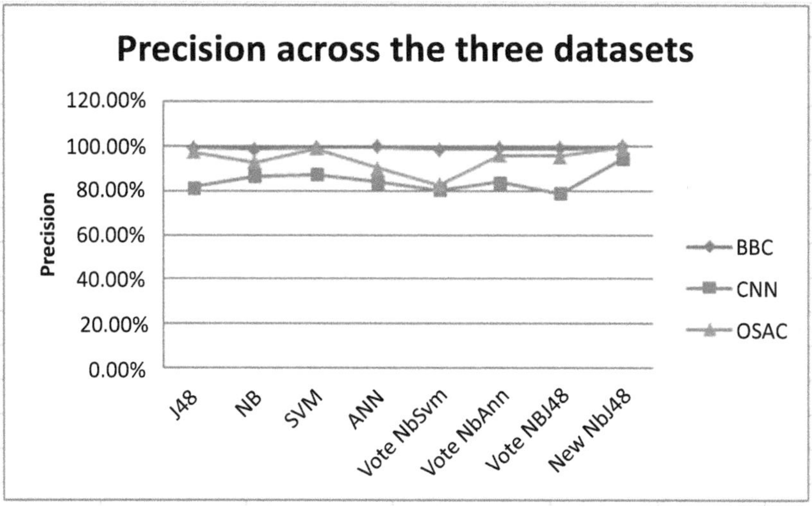

Figure 11.11 Precision across the three datasets.

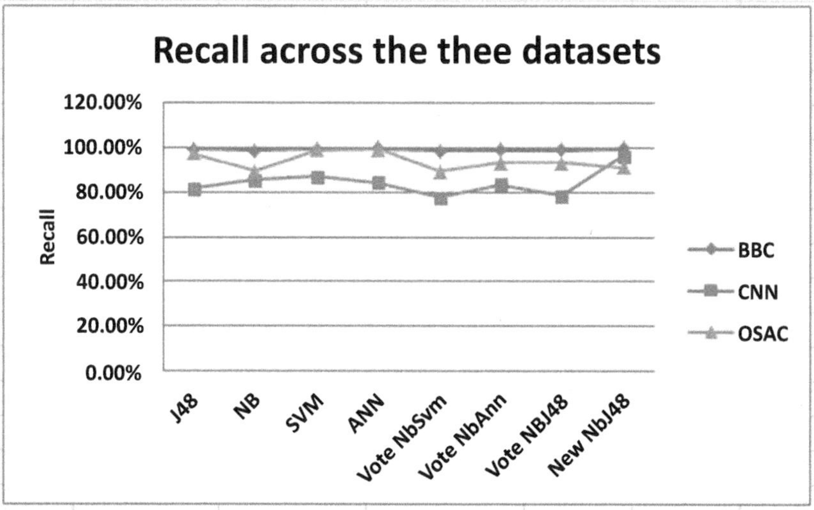

Figure 11.12 Recall across the three datasets.

11.5 Conclusion and future work

Text classification is a crucial requirement of the modern age, with endless applications it is a process that cannot be overlooked. When it comes to the Arabic language, the process of text classification becomes somewhat tricky due to the extreme complexity of the language.

In text classification, the naïve Bayes algorithm is widely and commonly used, therefore this thesis aimed at measuring the effect this common algorithm has on the process of text classification when it is combined with other commonly used algorithms. First, the algorithms were combined together within the WEKA tool platform using the vote method, and the results showed that the naïve Bayes algorithm degrades all of their performances. However, a new algorithm that has never been applied to Arabic text that also combined Naïve Bayes algorithm with another algorithm, namely J48, but using a different method than the voting method, showed drastic improvements and proved to be superior to all of the other algorithms whether combined or not combined with the naïve Bayes algorithm.

Finally, as a suggestion for future work, it is recommended to follow the new method of combining the naïve Bayes algorithm with other algorithms and measure the performance of these new hybrid algorithms.

References

[1] R. Feldman, J. Sanger, The Text Mining Handbook: Advanced Approaches in Analyzing Unstructured Data, Cambridge university press, 2007.
[2] C. Oppenheim, et al., The evaluation of WWW search engines, Journal of Documentation (2000).
[3] S.A. Kalogirou, Artificial neural networks in renewable energy systems applications: a review, Renewable and Sustainable Energy Reviews 5 (4) (2001) 373–401.
[4] A.P. Marugán, et al., A survey of artificial neural network in wind energy systems, Applied Energy 228 (2018) 1822–1836.
[5] D. Yousri, et al., Reliable applied objective for identifying simple and detailed photovoltaic models using modern metaheuristics: comparative study, Energy Conversion and Management 223 (2020) 113279.
[6] D.T. Hawkins, S.E. Larson, B.Q. Caton, Information science abstracts: tracking the literature of information science. Part 2: a new taxonomy for information science, Journal of the American Society for Information Science and Technology 54 (8) (2003) 771–781.
[7] K. Kowsari, et al., Text classification algorithms: a survey, Information 10 (4) (2019) 150.
[8] R. Elbarougy, G. Behery, A. El Khatib, Extractive Arabic text summarization using modified PageRank algorithm, Egyptian Informatics Journal 21 (2) (2020) 73–81.
[9] H.R. Pourghasemi, et al., Application of learning vector quantization and different machine learning techniques to assessing forest fire influence factors and spatial modelling, Environmental Research (2020) 109321.
[10] Q. Al-Radaideh, Applications of mining Arabic text: a review, Recent Trends in Computational Intelligence, IntechOpen, 2020.
[11] M. Scanagatta, A. Salmerón, F. Stella, A survey on Bayesian network structure learning from data, Progress in Artificial Intelligence (2019) 1–15.

[12] L. Abualigah, A. Diabat, Z.W. Geem, A comprehensive survey of the harmony search algorithm in clustering applications, Applied Sciences 10 (11) (2020) 3827.
[13] A. Purohit, et al., Text classification in data mining, International Journal of Scientific and Research Publications 5 (6) (2015) 1–7.
[14] S. Batra, S. Sachdeva, Organizing standardized electronic healthcare records data for mining, Health Policy and Technology 5 (3) (2016) 226–242.
[15] F. Elghannam, Text representation and classification based on bi-gram alphabet, Journal of King Saud University-Computer and Information Sciences (2019).
[16] Y. Liu, H.T. Loh, A. Sun, Imbalanced text classification: a term weighting approach, Expert Systems with Applications 36 (1) (2009) 690–701.
[17] L.M. Abualigah, et al., Text feature selection with a robust weight scheme and dynamic dimension reduction to text document clustering, Expert Systems with Applications 84 (2017) 24–36.
[18] L.M.Q. Abualigah, E.S. Hanandeh, Applying genetic algorithms to information retrieval using vector space model, International Journal of Computer Science, Engineering and Applications 5 (1) (2015) 19.
[19] I. Mokriš, L. Skovajsová, Neural network model of system for information retrieval from text documents in slovak language, Acta Electrotechnica et Informatica No 5 (1) (2005) 3.
[20] L.M. Abualigah, A.T. Khader, Unsupervised text feature selection technique based on hybrid particle swarm optimization algorithm with genetic operators for the text clustering, The Journal of Supercomputing 73 (11) (2017) 4773–4795.
[21] C. Gurrin, et al., A test collection for interactive lifelog retrieval, International Conference on Multimedia Modeling, Springer, 2019.
[22] L.M. Abualigah, A.T. Khader, E.S. Hanandeh, A new feature selection method to improve the document clustering using particle swarm optimization algorithm, Journal of Computational Science 25 (2018) 456–466.
[23] L.M.Q. Abualigah, Feature Selection and Enhanced Krill Herd Algorithm for Text Document Clustering, Springer, 2019.
[24] S.J. Cunningham, J. Littin, I.H. Witten, Applications of machine learning in information retrieval, 1997.
[25] L.M. Abualigah, A.T. Khader, M.A. Al-Betar, Unsupervised feature selection technique based on genetic algorithm for improving the text clustering, 2016 7th International Conference on Computer Science and Information Technology (CSIT), IEEE, 2016.
[26] O.A. Alomari, et al., Gene selection for cancer classification by combining minimum redundancy maximum relevancy and bat-inspired algorithm, International Journal of Data Mining and Bioinformatics 19 (1) (2017) 32–51.
[27] L. Mohammad Abualigah, et al., Hybrid harmony search algorithm to solve the feature selection for data mining applications, Recent Advances in Hybrid Metaheuristics for Data Clustering (2020) 19–37.
[28] L. Abualigah, et al., Sentiment analysis in healthcare: a brief review, Recent Advances in NLP: The Case of Arabic Language, Springer, 2020, pp. 129–141.
[29] L. Abualigah, et al., Text summarization: a brief review, Recent Advances in NLP: The Case of Arabic Language, Springer, 2020, pp. 1–15.
[30] M. Ikonomakis, S. Kotsiantis, V. Tampakas, Text classification using machine learning techniques, WSEAS Transactions on Computers 4 (8) (2005) 966–974.
[31] J. Singh, Challenges in text classification using machine learning techniques, International Journal of Recent Trends in Engineering and Research 4 (2) (2018) 81–83.

[32] O.A. Alomari, et al., MRMR BA: a hybrid gene selection algorithm for cancer classification, Journal of Theoretical and Applied Information Technology 95 (12) (2017) 2610−2618.
[33] F. Sebastiani, Machine learning in automated text categorization, ACM Computing Surveys (CSUR) 34 (1) (2002) 1−47.
[34] A.I. Kadhim, Survey on supervised machine learning techniques for automatic text classification, Artificial Intelligence Review 52 (1) (2019) 273−292.
[35] C.H. Wan, et al., A hybrid text classification approach with low dependency on parameter by integrating K-nearest neighbor and support vector machine, Expert Systems with Applications 39 (15) (2012) 11880−11888.
[36] L.M. Abualigah, A.T. Khader, E.S. Hanandeh, A novel weighting scheme applied to improve the text document clustering techniques, Innovative Computing, Optimization and Its Applications, Springer, 2018, pp. 305−320.
[37] S. Tan, An improved centroid classifier for text categorization, Expert Systems with Applications 35 (1−2) (2008) 279−285.
[38] J. Chen, et al., Feature selection for text classification with Naïve Bayes, Expert Systems with Applications 36 (3) (2009) 5432−5435.
[39] H. Sawaf, J. Zaplo, H. Ney, Statistical Classification Methods for Arabic News Articles, Natural Language Processing in ACL2001, Toulouse, France, 2001.
[40] R. Al-Shalabi, et al., Stop-word removal algorithm for Arabic language, Proceedings. 2004 International Conference on Information and Communication Technologies: From Theory to Applications, IEEE, 2004.
[41] F. Thabtah, et al., Naïve Bayesian based on Chi Square to categorize Arabic data, in: Proceedings of the 11th International Business Information Management Association Conference (IBIMA) Conference on Innovation and Knowledge Management in Twin Track Economies, Cairo, Egypt, 2009.
[42] F. Harrag, E. El-Qawasmeh, P. Pichappan, Improving Arabic text categorization using decision trees, 2009 First International Conference on Networked Digital Technologies, IEEE, 2009.
[43] R. Belkebir, A. Guessoum, A hybrid BSO-Chi2-SVM approach to Arabic text categorization, 2013 Acs International Conference on Computer Systems and Applications (AICCSA), IEEE, 2013.
[44] F. Harrag, E. El-Qawasmah, Neural network for Arabic text classification, 2009 Second International Conference on the Applications of Digital Information and Web Technologies, IEEE, 2009.
[45] S.A. Muqasqas, A Hybrid Classification Approach to Improve Classification Accuracy, Yarmouk University, 2013.
[46] R.M. Duwairi, Machine learning for Arabic text categorization, Journal of the American Society for Information Science and Technology 57 (8) (2006) 1005−1010.
[47] K. Darwish, Building a shallow Arabic morphological analyser in one day, in: Proceedings of the ACL-02 Workshop on Computational Approaches to Semitic Languages, 2002.
[48] M.K. Saad, W.M. Ashour, OSAC: Open Source Arabic Corpora, 2010, p. 10.
[49] L.M. Abualigah, A.T. Khader, E.S. Hanandeh, Hybrid clustering analysis using improved krill herd algorithm, Applied Intelligence 48 (11) (2018) 4047−4071.
[50] L.M. Abualigah, A.T. Khader, E.S. Hanandeh, A combination of objective functions and hybrid krill herd algorithm for text document clustering analysis, Engineering Applications of Artificial Intelligence 73 (2018) 111−125.
[51] A. Tripathy, A. Agrawal, S.K. Rath, Classification of sentiment reviews using n-gram machine learning approach, Expert Systems with Applications 57 (2016) 117−126.

[52] E. Frank, et al., Weka – a machine learning workbench for data mining, Data Mining and Knowledge Discovery Handbook, Springer, 2009, pp. 1269–1277.
[53] S. Vijayarani, S. Dhayanand, Liver disease prediction using SVM and Naïve Bayes algorithms, International Journal of Science, Engineering and Technology Research (IJSETR) 4 (4) (2015) 816–820.
[54] N. Guenther, M. Schonlau, Support vector machines, The Stata Journal 16 (4) (2016) 917–937.
[55] L.M. Abualigah, et al., A novel hybridization strategy for krill herd algorithm applied to clustering techniques, Applied Soft Computing 60 (2017) 423–435.

Index

Note: Page numbers followed by "*f*" and "*t*" refer to figures and tables, respectively.

Kolibri 360 GmbH
Ferdinand-Jühlke-Straße 7
99095 Erfurt, Deutschland
produktsicherheit@kolibri360.de